NATIONAL
GEOGRAPHIC

NATIONAL GEOGRAPHIC

TALES of the
unbelievable

weird
TRUE STORIES

Edited by
David Braun

NATIONAL GEOGRAPHIC
WASHINGTON, D.C.

Published by the National Geographic Society

Copyright © 2012 National Geographic Society

This 2014 edition printed for Barnes & Noble, Inc. by the National Geographic Society.

ISBN: 978-1-4351-5409-4 (B&N ed.)

Library of Congress Cataloging-in-Publication Data

National Geographic tales of the weird : unbelievable true stories / edited by David Braun.
  p. cm.
ISBN 978-1-4262-0965-9 (pbk. : alk. paper)
1. Curiosities and wonders. I. Braun, David. II. National Geographic Society (U.S.)
AG243.N38 2012
031.02--dc23

2012024608

The National Geographic Society is one of the world's largest nonprofit scientific and educational organizations. Founded in 1888 to "increase and diffuse geographic knowledge," the Society's mission is to inspire people to care about the planet. It reaches more than 400 million people worldwide each month through its official journal, *National Geographic,* and other magazines; National Geographic Channel; television documentaries; music; radio; films; books; DVDs; maps; exhibitions; live events; school publishing programs; interactive media; and merchandise. National Geographic has funded more than 10,000 scientific research, conservation and exploration projects and supports an education program promoting geographic literacy.

For more information, visit www.nationalgeographic.com.

National Geographic Society
1145 17th Street N.W.
Washington, D.C. 20036-4688 U.S.A.

For information about special discounts for bulk purchases, please contact
National Geographic Books Special Sales: ngspecsales@ngs.org
For rights or permissions inquiries,
please contact National Geographic Books
Subsidiary Rights: ngbookrights@ngs.org

Cover design by Jonathan Halling
Interior design by Melissa Farris
Printed in the United States of America
14/QGF-CML/1

Contents

You Can't Make This Stuff Up!

And to think it all began with a two-headed snake. A Spanish farmer captured one in 2002, and scientists were eager to study it. National Geographic Daily News published a story about the strange reptile on our website, and our readers went wild. More than a million people clicked on that weird tale, and then they came back for more and more stories.

Ten years and 8,000 stories later, more than 200 million individuals have clicked on our stories about strange and wonderful things. The two-headed snake was the first of many of the astounding National Geographic stories that lit up the Internet during the last decade. Our fans just can't get enough of tales about albino Cyclops sharks, fish with hands, zombie ants with mind-controlling fungi, top-secret photographs of Area 51, and the truth behind the Maya "Doomsday" calendar.

As the founding editor of National Geographic News, I have watched our community of fans grow from hundreds to millions to hundreds of millions. It's been a delight to publish stories that are as much fun to produce as they are to read. There isn't a day that goes by when the editors of National Geographic News do not find stories about new species, amazing animal secrets, the wonders of deep space—all weird discoveries that change our thinking about who we are and where we came from, the great enigmas of the universe.

Did comets make life on Earth possible? Will superhuman hearing soon be possible? Can stars "eat" other stars? Where was the world's oldest mattress found? These are only some of the hundreds of questions explored in our stories. The answers can be profound (and even disturbing), and they almost always lead to new questions.

National Geographic Tales of the Weird is our first reader filled with all kinds of these unbelievable true stories. National Geographic Books editor

Amy Briggs and I have selected some of the highlights of the first ten years of National Geographic News—the stories that were most popular with the National Geographic global audience as well as some of our personal favorites. From "Creepy Crawlies" to "Human History," each chapter is stuffed full of our strangest, oddest, and most truly fascinating stories.

As I picked each tale, I had to constantly remind myself that you can't make this stuff up. We are living in the real age of discovery, and I have observed that the world is more marvelous and mysterious than anything we can imagine. All we're doing is sharing this with millions of readers. It's got to be one of the best jobs in the world.

—David Braun
Editor in Chief, National Geographic Daily News
news.nationalgeographic.com

Contributors

The stories you're about to read were crafted by the talented journalists at National Geographic Daily News, a top-notch crew devoted to ferreting out the weird and wonderful facts about our world.

Carolyn Barry

David Braun

Anne Casselman

Ted Chamberlain

Charles Choi

Chris Combs

Christine Dell'Amore

Blake de Pastino

Willie Drye

Fritz Faerber

Brian Handwerk

Mason Inman

Victoria Jaggard

Sebastian John

Matt Kaplan

Rachel Kaufman

Lucas Laursen

Richard A. Lovett

Stefan Lovgren

Sean Markey

Hillary Mayell

Mati Milstein

Anne Minard

Dave Mosher

Paula Neely

Scott Norris

James Owen

Diana Parsell

Heather Pringle

John Roach

Ker Than

Traci Watson

Acknowledgments

Our thanks to the team at National Geographic Books who made this weird collection come together.

Amy Briggs, *Senior Editor*

Dee Wong, *Researcher and Writer*

Melissa Farris, *Art Director*

Ruthie Thompson, *Designer*

Rob Waymouth, *Illustrations Editor*

Marshall Kiker, *Associate Managing Editor*

Judith Klein, *Production Editor*

Lisa A. Walker, *Production Manager*

Galen Young, *Rights Clearance Specialist*

Katie Olsen, *Design Assistant*

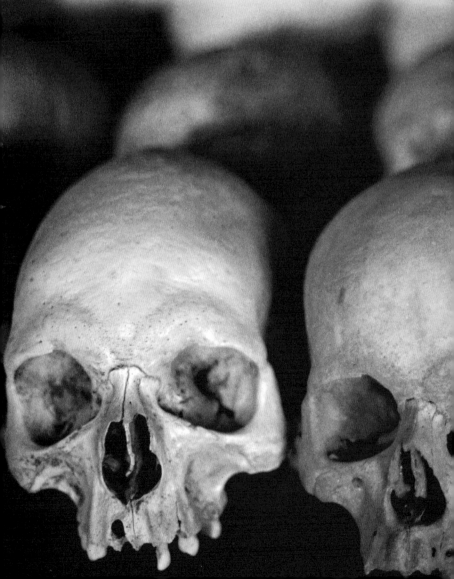

CHAPTER 1

Ancient
Sacred

Rites and Places

For centuries, human civilizations have been grappling with some very big spiritual issues: What happens when we die? How will the world end? Will I need my chariot in the afterlife? Strange and wonderful archaeological discoveries—from the puppy mummies of ancient Egypt to the entrance to the Maya underworld in Mexico—are revealing the many fascinating ways that cultures all over the world developed sacred rituals and practices.

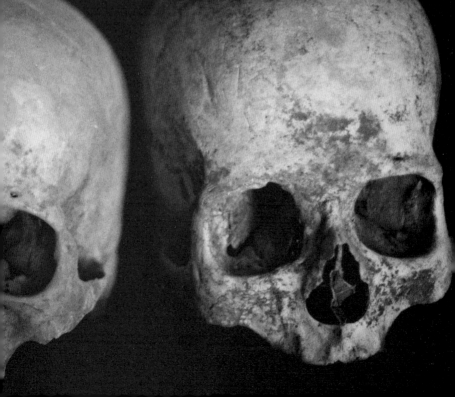

Maya "Doomsday"
Calendar Explained

It's remotely possible the world will end in December 2012. But don't credit the ancient Maya calendar for predicting it, say experts on the Mesoamerican culture.

It's true that the so-called long-count calendar—which spans roughly 5,125 years starting in 3114 B.C.—reaches the end of a cycle on December 21, 2012. That day brings to a close the 13th Bak'tun, an almost 400-year period in the Maya long-count calendar.

But rather than moving to the next Bak'tun, the calendar will reset at the end of the 13th cycle, akin to the way a 1960s automobile would click over at mile 99,999.9 and reset to zero.

FAILED PROPHECIES

70: Ancient Romans believe the end is near with the eruption of Mount Vesuvius.

1666: The Great Fire of London, along with the biblical denouncement of the number 666, contributes to Londoners' belief that this is their final year.

1910: The appearance of Halley's comet stirs up apocalyptic fears among Europeans and Americans, who believe gases in the comet's tail will end life on Earth.

1914: Since its founding in the 1870s, Jehovah's Witnesses predicted the end of the world in 1914. When it didn't come, the religion's followers began predicting that the end is coming "shortly."

"We, of course, know that really means a hundred thousand [miles] and not zero," said William Saturno, an expert on Maya archaeology at Boston University and National Geographic explorer.

"So, is [the end of Bak'tun 13] a large period ending? Yes. Did the Maya like period endings? Yes," Saturno said.

"Would this have been a period ending they thought was wicked cool? You bet. The biggest period endings they experience are Bak'tun endings."

But "was it predicted to be the end of the world? No. That's just us."

Instead, for the Maya, the end of the long count represents the end of an old cycle and the beginning of a new one, according to Emiliano Gallaga Murrieta, the Chiapas state division director of Mexico's National Institute of Anthropology and History.

> **"Two women in the last two weeks said they were contemplating killing their children and themselves so they wouldn't have to suffer through the end of the world."**
> **David Morrison**
> *senior scientist*
> *NASA Astrobiology Institute*

"It is like for the Chinese, this is the Year of the [Rabbit], and the next year is going to be the Year of the Dragon, and the next is going to be another animal in the calendar," Murrieta said.

Bak'tun the 13th

Written references to the end of Bak'tun 13 are few. In fact, most Maya scholars cite only one: a stone tablet on Monument 6 at the Tortuguero archaeological site in Mexico's Tabasco state.

What exactly the tablet says, though, is a mystery, because the glyphs in question are partially damaged. Nevertheless, scholars have taken several

March 1997: Heaven's Gate members commit suicide when comet Hale-Bopp is closest to Earth because they believe a UFO riding the comet will save them from the Apocalypse.

January 1, 2000: A 1984 trade publication predicts that computers will be crippled by a "Y2K" bug and cause mass chaos.

May 5, 2000: Richard Noone predicts that the planetary alignment of Mercury, Venus, Mars, Jupiter, and Saturn with the sun and the moon will cause another ice age to occur.

September 2009: Critics of the Large Hadron Collider, the world's largest atom smasher, believe that it will create a black hole that destroys Earth.

Maya pyramid of Kukulcan, Chichén Itzá, Mexico

stabs at translations, the most prominent in 1996 by Brown University's Stephen Houston and the University of Texas at Austin's David Stuart.

Houston and Stuart's initial interpretation indicated that a god will descend at the end of Bak'tun 13. What would happen next is uncertain, although the scholars suggested this might have been a prophecy of some sort.

This 1996 analysis was picked up "on many New Age websites, associated forum discussions, and even a few book chapters" as evidence that the Maya calendar had predicted the end of the world, according to Stuart.

Commemorating the Future

Houston and Stuart, however, recently revisited the glyphs independently and concluded that the inscription may actually contain no prophetic statements about 2012 at all.

TRUTH:
THE MAYA LIVED IN THE JUNGLES OF THE AMERICAS FOR 3,000 YEARS.

Rather, the mention of the end of Bak'tun 13 is likely a forward-looking statement that refers back to the main subject of the inscription, which is the dedication of Monument 6.

In a blog post about his conclusions, Stuart makes an analogy to a scribe wanting to immortalize the New York Yankees' 1950 sweep of the Philadelphia Phillies in that year's World Series.

If this writer were to use the Maya rhetorical device thought to be in Monument 6's inscription, the text might read as follows:

On October 7, 1950, the New York Yankees defeated the Philadelphia Phillies to win the World Series. It happened 29 years after the first Yankees victory in the World Series in 1921. And so 50 years before the year 2000 will occur, the Yankees won the World Series.

Written this way, Stuart notes, the text mentions a future time of historical importance—the 50-year anniversary of the victory—but it does so in reference to the event at hand, i.e., the 1950 game.

"This is precisely how many ancient Maya texts are structured, including Tortuguero's Monument 6," Stuart writes.

Pure Poetry?

According to INAH's Murrieta, this structure of Maya texts is what has confused modern minds, given our penchant for literal, straightforward reading. Even if the Monument 6 inscription refers to a god coming down at the end of Bak'tun 13, it isn't a statement about the end of the world, he said.

"They are writing in a more poetic sense, saying, Well, on the 21st of December 2012, the god is going to come down and start a new cycle and the old world is going to die and the new world is going to be reborn—just to make it more poetic."

Saturno, the Boston University archaeologist, agreed that the reference to a specific date is clear in Monument 6, but added that "there's no text that follows and says, Herein will be the end of the world, and the world will end in fire...That's not anywhere in the text."

Rather, Saturno said, the hype around 2012 stems from dissatisfied Westerners looking to the ancients for guidance, hoping that peoples such as the Maya knew something then that could help us through difficult times now.

In any case, even if the ancient inscriptions explicitly predicted the end of the world, Saturno wouldn't be worried, given the Maya track record with long-range prophecy.

"They didn't see [their] collapse coming. They didn't see the Spanish conquest coming." ∎

> "Our ancestors said that when the last days draw closer, many people will die and bad things will happen."
> **Mary Coba Cupul**
> *Maya descendant*

Millions of Puppy Mummies

As part of the first full excavation of Egypt's ancient Dog Catacombs, scientists are examining 2,500-year-old animal remains—a small sample of the roughly eight million animal mummies in these tunnels.

Snaking beneath the desert at the ancient royal burial ground of Saqqara, the Dog Catacombs were discovered more than a century ago. But only now is research shedding light on the massive number of mummies found in this complex of tunnels and chambers dedicated to Anubis, the jackal-headed god of the afterlife. Poorly mummified and piled high, the carcasses long ago deteriorated into indistinct heaps, experts say. The mummies were stacked between about the late sixth century B.C. and the late first century B.C.

"It's not easy to identify individual mummies in the galleries or in the photographs," said Paul Nicholson of Cardiff University in the U.K. "We have piles of mummy remains just over a meter [three feet] high, on average, that just fill the side tunnels.

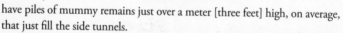

"Although the mummies are not well preserved or well decorated, unlike some museum specimens, they can still give us a great deal of scientific information."

From Your Mouth to Dog God's Ears

Piled with decomposing animal mummies, the tunnels in the Dog Catacombs are evidence of ancient Egyptian pilgrims' fierce desire to be heard by the canine-headed god Anubis. Today, "in some churches people light a candle, and their prayer is taken directly up to God in that smoke," archaeologist Salima Ikram said.

In much the same way, a mummified dog's spirit would carry a person's prayer to the afterlife, said Ikram, founder of the Animal Mummy Project at the Egyptian Museum in Cairo. "And because it's the same kind of species as the god, it would have special access to Anubis's ear. So the person would have a sort of direct line to the god," Ikram added.

Bodies of Evidence

Mummified dogs such as this aren't the only animals filling the Dog Catacombs. "There are also some jackals, foxes, and ichneumon [Egyptian mongooses]," said Cardiff University's Nicholson. "But the overwhelming majority of animals here are dogs, so we believe this place was intended just for the cult of Anubis."

Elsewhere in Saqqara's so-called Sacred Animal Necropolis, there were many other cults—and many other mummified animals, including baboons, cows, bulls, and so on, said Nicholson, whose work was partly funded by the National Geographic Society's Committee for Research and Exploration.

> "There is also still debate about exactly what kind of canid Anubis was meant to represent . . . We hope to discover more information about the . . . animals in the catacombs to understand how the god was perceived."
> **Paul Nicholson**
> director, Cardiff University's Egypt Exploration Society Mission to the Dog Catacombs

All Shapes and Sizes

The mazelike Dog Catacombs hold dogs of "all shapes and sizes, all ages, from fetal to mature," the Egyptian Museum's Ikram said. "Some are short-legged, more like dachshund types, while others are long-legged and more

Mummified dog remains from the Dog Catacombs

like golden retriever types. I think they [mummified] whatever was there at the time and whatever suited the pilgrim's pocket."

Ancient puppy farms likely operated in the area to supply dogs in such numbers, and many of the animals were mummified as newborns, Cardiff's Nicholson said. "Perhaps the priests of Anubis would have regularly taken a certain number of animals each week to mummify for pilgrims."

There are mummified dogs that occupy special niches in the catacombs—places likely commensurate with an honored role in real life. A number of older, male dogs that might have lived in the nearby Temple of Anubis were mummified with far more care than the rest, Ikram speculated. "We think that these might have been

TRUTH: THE FIRST TOMBS WERE BUILT TO PROTECT THE DEAD FROM ROAMING DOGS AND JACKALS.

the actual sacred dogs that were the manifestation of Anubis on Earth," she added.

With the permission of Egypt's Ministry of Antiquities and in association with the Egypt Exploration Society, archaeologists continue to sift sands at the Dog Catacomb. Nicholson said, "For the first time at this site, we're looking at a really large sample of the dog mummies and trying to learn how old they are, what gender they are, whether we can say anything about their species or how they met their deaths."

Other Animal Occupants

Strangely, a small number of cat mummies have also been found in the Dog Catacombs. Perhaps, Ikram speculated, the felines had taken up residence in the Temple of Anubis and become sacred by association. Or maybe they were fakes. "There may have been some time when they didn't have enough dogs and decided to just wrap up a cat and make it look like a dog," she said. "It might have been a charlatan moment."

Holy Animals

Dogs, foxes, and jackals were sacrificed as offerings to the god Anubis, but they weren't the only ones. There were a number of animals that were the focus of worship for particular gods in ancient Egyptian society. Major changes that happened between 3000 and 2000 B.C. elevated these animals to sacred status. Here were some of the most popular gods and their corresponding animal cults:

- Apis—Bull
- Bastet—Cat
- Hathor—Cow
- Horus—Hawk or falcon
- Khnum—Ram
- Sobek—Crocodile
- Thoth—Baboon or ibis

Among other remaining mysteries are oddly avian mummies. "They are wrapped up as falcons," Ikram said, "but until we're able to have them x-rayed, we won't know exactly what they are." ■

Lifelike Chinese "Wet Mummy"

Construction on a new highway yielded a remarkable find: a 600-year-old Chinese mummy with her eyebrows, hair, and skin still intact. Her identity remains a mystery.

Ice Mummy

In 1991, climbers discovered a frozen body on a glacier near the Austrian-Italian border. At first, forensic experts didn't realize how old the body was. Using radiocarbon dating, it was later determined that this "iceman" died at some point between 3350 and 3300 B.C., which makes him the oldest well-preserved mummy in the world.

A fortunate series of events probably led to the remarkable preservation of a 600-year-old Chinese woman whose tomb had been discovered accidentally by road builders near the city of Taizhou, China. "Wet mummies survive so well because of the anaerobic conditions of their burials," said archaeologist Victor Mair. That is, water unusually void of oxygen inhibits bacteria that would normally break down a body.

Unlike ancient Egyptian mummies, the corpse—likely from the Ming dynasty (1368 to 1644)—was probably preserved only accidentally, said Mair, of the University of Pennsylvania. "I don't know of any evidence that Chinese ever intentionally mummified their deceased," he said. "Whoever happened to encounter the right environment might become a preserved corpse."

Who Are You?

When staff members from China's Taizhou Museum carefully removed the woman's mummy—one of three found during a road expansion—from her wooden coffin, they used the items found with her to tell the story of her life. Adorning her chest was a so-called exorcism coin. "My guess would be that the coin was placed on the body as a kind of charm against malevolent influences," said Timothy Brook, a historian at the University of British Columbia's Institute of Asian Research.

Her fully dressed, 5-foot-long (1.5-meter-long) body was buried with luxury items, including a jade ring, a silver hairpin, and more than 20 pieces of Ming-dynasty clothing. Jade was associated with longevity in ancient China. But in this case, the jade ring was "probably a sign of her wealth instead of a sign of any concern about the afterlife," Brook said.

The lack of identifying insignia such as a phoenix or dragon, though, suggests the wet mummy wasn't royal, said Brook. "Her headgear is sort of ordinary," Brook observed. "There's nothing that sets her apart from anyone else . . . She was probably just a well-off person."

Still unknown is whether the woman was buried with any written documents or inscribed pottery—a common practice in Ming-dynasty China. "If you were a person of any importance, you had someone write a [remembrance] or a brief biography," Brook said, "and that biography would often be posted at the burial site and a copy buried with you as well, to identify who you are" in the afterlife.

Good Hair Day

When museum staff removed the female mummy's cap, her head appeared to be dyed a bright shade of purple. Neither Brook nor the University of Pennsylvania's Mair know exactly why her head is that color, though Mair speculates it may have to do with natural minerals in the water.

The mummy's hair was still held in place by a bright silver hairpin—a fairly standard example of the type worn by Ming-dynasty women, Brook said. The woman's exact age at death is unknown, but her unlined face suggests she was fairly young. "She's certainly an adult," Brook said. "She's not an old woman."

TRUTH:
THE WORD "MUMMY" DERIVES FROM THE PERSIAN AND ARABIC WORDS "MUM" AND "MUMYA," WHICH ARE USED TO DESCRIBE WAX OR BITUMEN.

A Positive Judgment

In ancient China, it was believed that the newly dead would appear before supernatural judges. Brook explained that "If you were found to be morally worthy you would be sent off for reincarnation—as a deity if you were fantastic, as a human being if you were good, as an animal if you were less good, and as a bug or a worm if you'd really bad."

During the Ming dynasty, preservation after death was thought to "reflect your purity" in life, Brook explained. Had this woman's family known her body would be preserved for more than 600 years, they would have been extremely proud, he added. ■

Celtic Princess Tomb

Yields Gold, Amber Riches

An 80-ton treasure trove of a tomb has a new home near Stuttgart, where it was moved after being completely lifted out of the ground and hauled away.

Discovered in December 2010 beside the River Danube in Heuneburg, a Celtic princess's tomb—all 80 tons of it—was lifted whole by heavy cranes and transported to a tented laboratory outside of Stuttgart, Germany, where archaeologists with the Stuttgart Regional Council are now analyzing the contents, many of which are in amazing condition.

The excavation of the ancient Celtic burial chamber, August 2010

DIGGING FOR TREASURE

1: Two heavy cranes lift the **80-ton tomb** from the frozen earth in December 2010 in preparation for transport to an archaeological facility near Stuttgart.

2: Finely worked **gold and amber jewelry,** including these pieces, are among scores of Celtic treasures from a tomb recently unearthed in southern Germany.

3: Undetected for more than two millennia, the **ancient Celtic burial chamber** begins to reveal its secrets in a picture taken at the dig site in November 2010.

1

2

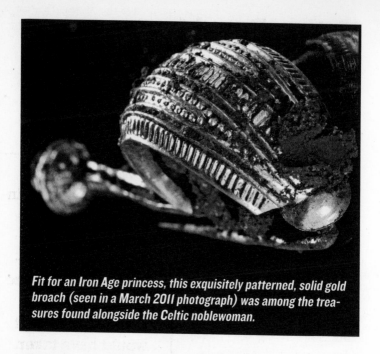

Fit for an Iron Age princess, this exquisitely patterned, solid gold broach (seen in a March 2011 photograph) was among the treasures found alongside the Celtic noblewoman.

Perfect for a Princess

The well-preserved, large wooden burial chamber contains the 2,600-year-old skeleton of an ancient Celtic noblewoman. Aged between 30 and 40 when she died, the highborn lady was buried with a cache of ornate treasures, such as gold necklaces set with pearls, and she was found wearing crafted amber around her waist. The lavish grave confirms Heuneburg as one of the earliest centers of Celtic art and culture, according to excavation leaders.

"Normally these graves are robbed over the centuries, so it's amazing to have one in such good condition," said project co-leader Nicole Ebinger-Rist, an archaeologist with Stuttgart Regional Council. Treasures inside the tomb include engraved copper-alloy plates and gold-encrusted animal teeth, which the Celts used for adorning their horses, she added.

TRUTH: A GOLD NUGGET FOUND IN CALIFORNIA WEIGHED 160 POUNDS—ABOUT AS MUCH AS 12 BOWLING BALLS.

Mother-Daughter Relationship?

The woman's tomb was found not far from the similarly aged grave of a girl, previously unearthed in 2005, whose body was adorned almost identically, Ebinger-Rist said.

"What's amazing is that the jewelry from the girl—two brooches and two earrings—have exactly the same type of ornamentation," she said. "These must be two related people."

The study team hopes DNA analysis will reveal whether the pair belonged to the same family—or were even mother and daughter.

So Well Preserved

The unusual preservation of the oak floor—caused by waterlogging of the 8-meter-long (26-foot-long) burial chamber—was key to providing an accurate date for the site, Ebinger-Rist said. Analysis of the wood indicates that oak trees used for the chamber's construction were felled 2,620 years ago, making this royal Celtic grave one of the earliest ones known.

Supported by 14 large steel pipes and a surrounding steel frame, the burial chamber was moved so that researchers could sift through the find in microscopic detail under laboratory conditions. Already the lab team has recovered minute fragments of leather and clothing and traces of corroded copper alloy plates covered in engravings, according to Ebinger-Rist. "This is amazing—out in the field we have no chance of detecting such objects," she said. ■

> "We found fabulous leather belts in some of the high-status women's graves, with thousands of tiny bronze staples attached to the leather that would have taken hours to make ... I call them the Iron-Age Harley-Davidson biker chicks."
> **Bettina Arnold**
> *archaeologist and co-director of a field excavation at the Heuneburg hillfort*

Ancient Sorcerer's "Wake"

First Feast for the Dead?

Evidence found in the resting place of an ancient shaman indicates that perhaps the world's first villagers fostered peace via partying.

Some 12,000 years ago in a small sunlit cave in northern Israel, mourners finished the last of the roasted tortoise meat and gathered up dozens of the blackened shells. Kneeling down beside an open grave in the cave floor, they paid their last respects to the elderly dead woman curled within, preparing her for a spiritual journey.

Tortoise shells surround the shaman's body.

They tucked tortoise shells under her head and hips and arranged dozens of the shells on top and around her. Then they left her many rare and magical things—the wing of a golden eagle, the pelvis of a leopard, and the severed foot of a human being.

A Spiritual Site

Now called Hilazon Tachtit, the small cave chosen as this woman's resting place is the subject of an intense investigation led by Leore Grosman, an archaeologist at the Hebrew University of Jerusalem in Israel.

Already her research has revealed that the mystery woman—a member of the Natufian culture, which flourished between 15,000 and 11,600 years ago in what is now Israel, Jordan, Lebanon, and possibly Syria—was the world's earliest known shaman. Considered a skilled sorcerer and healer, she was likely seen as a conduit to the spirit world, communicating with supernatural powers on behalf of her community, Grosman said.

A study published in the journal *Proceedings of the National Academy of Sciences* by Grosman and Natalie Munro, a zooarchaeologist at the University of Connecticut, reveals that the shaman's burial feast was just one chapter in the intense ritual life of the Natufians, the first known people on Earth to give up nomadic living and settle in villages.

> "From the standpoint of the status of the grave and its contents, no Natufian burial like this one has ever been found. This indicates the woman had a distinct societal position."
>
> **Leore Grosman**
> *archaeologist,*
> **Hebrew University of Jerusalem**

First Feasts for the Dead

In the years that followed the burial, many people repeatedly climbed the steep, 492-foot-high (150-meter-high) escarpment to the cave, carrying up other members of the community for burial as well as hauling large amounts of food. Next to the graves, the living dined lavishly on the meat of aurochs, the wild ancestors of cattle, during feasts conducted perhaps to memorialize the dead.

New evidence from Hilazon Tachtit, in northern Israel's Galilee region, suggests that mortuary feasting began at least 12,000 years ago, near the end of the Paleolithic era. These events set the stage for later and much more elaborate ceremonies to commemorate the dead among Neolithic farming communities.

In Britain, for example, Neolithic farmers slaughtered succulent young pigs 5,100 years ago at the site of Durrington Walls, near Stonehenge, for an annual midwinter feast. As part of the celebrations, participants are thought to have cast the ashes of compatriots who had died during the previous year into the nearby River Avon.

Just This Once ...

For the burial wake of the shaman, the Natufian people feasted on aurochs and wild tortoise, but their day-to-day diet was less extravagant. As a result of transitioning from a nomadic, hunter-gatherer culture to a sedentary, agriculture-based lifestyle, the food they ate reflected both ways of life and included wild cereals, legumes, almonds, acorns, gazelle, deer, beef, wild boar, ibex, duck, and fish, depending on the area of settlement.

The Natufian findings give us our first clear look at the shadowy beginnings of such feasts, said Ofer Bar-Yosef, an archaeologist at Harvard University. "The Natufians," Bar-Yosef said, "were like the founding fathers, and in this sense Hilazon Tachtit gives us some of the other roots of Neolithic society." Study co-author Grosman agrees. "The Natufians," she said, "had one leg in the Paleolithic and one leg in the Neolithic."

Exploring the Cave

Perched high above the Hilazon River in western Galilee, Hilazon Tachtit cave was long known only to local goatherds and their families. But in the early 1990s, Harvard's Bar-Yosef spotted several Natufian flint artifacts scattered along the arid, shrubby slope below the cave and climbed up to investigate.

Impressed by the site's potential, the Harvard University archaeologist recruited Hebrew University's Grosman to take charge of the dig, and she and a small team began excavations there in 1995.

First Grosman and her team had to peel back an upper layer of goat dung, ash, and pottery shards that had accumulated during the past 1,700 years. Below this layer they found five ancient pits filled with bones, distinctive Natufian stone tools, and pieces of charcoal that dated the pits to between 12,400 and 12,000 years ago.

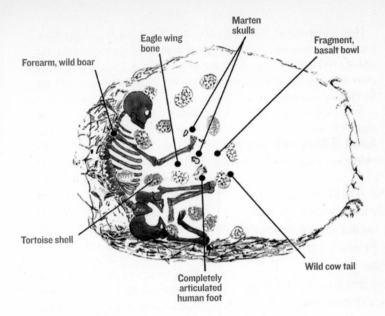

Marten skulls

Eagle wing bone

Fragment, basalt bowl

Forearm, wild boar

Tortoise shell

Wild cow tail

Completely
articulated
human foot

The Shaman's Body

At the bottom of one pit lay the 45-year-old shaman—quite elderly for Natufian times—buried with at least 70 tortoise shells and parts of several rare animals. Analyses showed that this woman had suffered from a deformed pelvis. She would have had a strikingly asymmetrical appearance and likely limped, dragging her foot.

Grosman examined historical accounts of shamans worldwide and found that in many cultures shamans often possessed physical handicaps or had suffered from some form of trauma. According to Brian Hayden, an archaeologist at Simon Fraser University in Burnaby, Canada, "It's not uncommon that people with disabilities, either mental or physical, are thought to have unusual supernatural powers."

Opened Graves

After burying their spiritual leader 12,000 years ago at Hilazon Tachtit, Natufians returned to the cave for other funerary rituals, eventually interring the bodies of at least 27 men, women, and children in three communal burial pits, researchers say.

On some later visits, Natufians opened the communal graves and removed certain bones, including skulls, for possible display or burial elsewhere, according to Grosman.

Until now, removing bones from burials for use in rituals was thought to have begun during the Neolithic era at sites such as the West Bank's Jericho, dating to about 11,000 years ago. A similar practice has been found at the later Neolithic site of Çatalhöyük in Turkey. In both places mourners coated human heads with plaster and kept them for ceremonial purposes.

On the Menu

"We think that there were scheduled visits to Hilazon Tachtit," said study co-author Grosman, who received partial funding from the National Geographic Society's Committee for Research and Exploration for her work at the Natufian site.

Natufians seem to have made the steep climb laden with joints of mountain gazelle and aurochs. In what might have been one sitting, the mourners devoured an estimated 661 pounds (300 kilograms) of aurochs meat, according to the study.

But they did not bring all this food for a picnic. In their daily lives, Natufian families seldom dined on aurochs, for the wild oxen were relatively scarce at this time. And given the species' power and speed, hunting an aurochs likely required a communal effort.

The celebrants chose to feast on aurochs for reasons above and beyond their nutritional value. "In later times we know that the aurochs become ritually important in the area," zooarchaeologist Natalie Munro said. Indeed, some later cultures seem to have regarded aurochs as sacred animals, even symbols of fertility.

For example, at the massive, 11,600-year-old Gobekli Tepe ritual site in Turkey—seen by some as the world's oldest temple—hunters and gatherers dined lavishly on aurochs.

> **Buried Treasure**
>
> Among the artifacts buried with the Natufian shaman:
>
> 1. Tortoise shells
> 2. Leopard pelvis
> 3. Complete articulated human foot
> 4. Wild cow tail
> 5. Fragment of a basalt bowl
> 6. Marten skulls
> 7. Eagle wing bone
> 8. Forearm of a wild boar

Good Food Makes Good Neighbors

Munro thinks that the grand ritual feasts at Hilazon Tachtit served an important purpose besides mourning lost loved ones. Living for the first time in settled communities, Natufian families had to find a way to ease all the friction that would build up from continually rubbing shoulders with their neighbors, she says. Unlike other Paleolithic hunters and gathers, the Natufians could no longer split up and move on easily when trouble arose. They had become so populous that they could no longer find unoccupied territory in their region.

The Natufians devised a way of dealing with the strain—throwing big communal parties to celebrate important ritual events, the researchers say. "When people feel like they are part of the same group, they are more willing to share and to compromise to resolve conflict," Munro said.

TRUTH:
THE WORD "SHAMAN" WAS COINED IN SIBERIA, BUT THE CONCEPT IS COMMON IN CULTURES AROUND THE WORLD.

The new finds suggest that the deep roots of communal feasting and the curation of human remains for ritual—found later at sites like Gobekli Tepe, Jericho, and Stonehenge—originated centuries before the advent of agricultural societies.

These rituals played an important role in smoothing the transition of Paleolithic hunter-gatherers to a farming life, researchers say. "Hilazon Tachtit," archaeologist Ofer Bar-Yosef said, "gives us a good window into these kinds of special activities. And I think that's really important." ∎

Ancient Chinese Chariot Fleet

Uncovered in a tomb pit, these millennia-old chariots and horse skeletons may have been part of an elaborate funeral rite for a wealthy nobleman.

Five chariots and 12 horse skeletons were found in a tomb pit unearthed in the city of Luoyang in central China. Archaeologists believe the tomb was dug as part of the funeral rites of a minister or other nobleman during the Eastern Zhou dynasty period, about 2,500 years ago.

Got Wheels?

Chariots were important vehicles of war during the Zhou dynasty and were driven by nobleman-warriors wielding halberds or spears, said David Sena, a China historian at the University of Texas at Austin who was not involved in the discovery.

War Machine

The chariot was the world's first war machine, originally used in Mesopotamia and then in Asia Minor and Egypt. It was an early version of a tank, from which arrows were shot as it raced through the battlefield.

"During this period, there wasn't a distinction between the military class and an educated aristocratic class," Sena said. "People with aristocratic backgrounds were expected to do both, and riding a chariot was one of the skills that a nobleman was expected to have."

"Chariots were kind of the main units of warfare during" the Zhou dynasty, he added. "Later, chariots were still used, but their effectiveness was reduced, because you got larger armies composed mostly of commoners, so warfare became much less of an aristocratic affair."

A worker sprays water onto some of the recently unearthed chariot horse skeletons to help them retain moisture.

Tricked-Out Chariots

Many chariots used by noblemen during the Eastern Zhou dynasty were adorned with valuable metals and materials, such as bronze and ivory, though reports of the Luoyang find make no mention of precious materials. Valuable chariot parts and accessories were often gifts from the Zhou dynasty's ruler himself, Sena explained.

"Many bronze inscriptions describe a kind of political ritual where a nobleman is invested with a title or duty or some land, and that's always accompanied by gift giving. Chariot parts and accoutrements were a very important part of that."

The Story in the Tomb

Archaeologists believe the 12 deliberately arranged horses were slaughtered prior to being buried. Arranged with the horses' bodies were several dogs. Dogs performed important work and were sacrificial animals in ancient China, so their presence in the chariot tombs is not unusual, Sena said. "We often see them in the bottom of pits of human tombs" as well, he added.

It's unclear whether the chariots and horses found in the recently excavated pit were expected to be of use to the buried aristocrat in the afterlife or whether the practice underscored a family's importance and wealth in life.

It may have been a combination of both, Sena said. "I think we can make the inference that these [chariots and sacrificed horses] spoke to some need that the dead would have in the afterlife."

Luoyang is currently undergoing rapid expansion, so "archaeologists are really kind of under the gun to save as much of this material before it's destroyed or covered over by industrial growth," Sena said. "It's very nice to see that they're able to save these things and document them for scholars and for posterity."

SWEET RIDE!

1: The bodies of **two horses and a dog** were carefully arranged in a smaller grave site discovered at the same time as the larger chariot tomb.

2: To protect it from drying out, a worker sprays water onto a **millennia-old chariot** recently unearthed in the city of Luoyang in central China.

3: The **newly unearthed chariot tomb** dates to the "heyday of chariot warfare" in China. ∎

1

2

3

New Death Ritual
Found in Himalaya

**Found in cliffside caves in Nepal,
the remains of 27 ancient men, women,
and children bear cut marks that
point to a previously unknown
Himalayan death ritual, experts say.**

The newly discovered corpses—many of which had been stripped of flesh—were placed in the high mortuaries some 1,500 years ago. Nearly 67 percent of the bodies had been de-fleshed, most likely with a metal knife, say the researchers, who found the remains in 2010.

After the de-fleshing process, the corpses had been neatly laid to rest on wide wooden shelves, the researchers speculate. But due to centuries of exposure to the elements, the bones and bunks—and much of the caves themselves—had collapsed by the time the team entered the chambers. Also in the jumble: goat, cow, and horse remains—perhaps sacrificial offerings for the dead, though their purpose remains a mystery.

> # TRUTH:
> ## THE HIMALAYAN MOUNTAINS GROW A HALF INCH TALLER EACH YEAR.

Cliff Caves

Dug into characteristically reddish cliffs of the Upper Mustang district, the human-made caves lie at 13,800 feet (4,200 meters) above sea level, high above the village of Samdzong.

A climber removes a skull from a cave in Nepal's Mustang region.

In ancient times, rock outcrops and probably ladders would have eased access to the caves. Since then, however, erosion has rendered the chambers accessible to only expert climbers, such as seven-time Everest summiteer Pete Athans, who co-led the team.

"Clues to when these caves were built, and by whom, are melting before our eyes," Athans said in a press statement. "The cave tomb we found is under great threat. It is situated in a fragile rock matrix that has already collapsed some time in the past. I don't believe the tomb would've lasted one more monsoon."

Treated With Respect

Little is known about the three ancient Himalayan groups that de-fleshed and entombed their dead in the high Mustang caves, making the motives behind the rite even murkier. The team has, however, ruled out cannibalism.

"When you're going for meat, you process a skeleton in a very different way than if you were trying to strip the flesh off," explained project leader Mark Aldenderfer, an archaeologist at the University of California, Merced.

"In cannibalism, the base of the skull is often smashed [to get at the brains], and bones are broken and twisted, usually for marrow. There's nothing like that in any of the bone parts that we recovered . . .

> **TRUTH:**
> **A CAVE IN CROATIA HAS A 1,683-FOOT-DEEP PIT, THE DEEPEST HOLE ON EARTH.**

"This was done in a respectful fashion," added Aldenderfer, who received partial funding from the National Geographic Society's Committee for Research and Exploration.

Preliminary DNA analysis of some of the bones suggests the de-fleshing subjects were related. "I would imagine that many of these mortuary caves are for large extended families," Aldenderfer said.

"This would be their traditional burial place, and another family would have their own."

Secondhand Rite?

Aldenderfer and his team think the practice of de-fleshing corpses and entombing them in caves might be a previously unknown bridge between two other known death rituals.

DEATH RITUALS FROM AROUND THE WORLD

AIR SACRIFICE, MONGOLIA	SKY BURIAL, TIBET	PIT BURIAL, PACIFIC NORTHWEST HAIDA	VIKING BURIAL, SCANDINAVIA
After a traditional ceremony, the body is laid on open ground and village dogs are released to consume the remains.	The deceased is dismembered by a *rogyapa*, or body breaker, and left outside to the natural elements.	The indigenous people of the American northwest coast would place their dead into a large open pit behind the village.	The Vikings were buried in large graves dug in the shape of a ship and lined with rocks.

One, the Tibetan sky burial—thought to have originated several hundred years later—involves dismembering a body and exposing it to the elements and to scavengers such as vultures. Present-day Tibet is just 10 miles (16 kilometers) from the cave tombs.

The other funerary rite is older and hails from the Zoroastrian religion, which has its roots in ancient Persia (now Iran). Zoroastrians, Aldenderfer said, "are known to have de-fleshed their dead and fed the flesh to animals."

Ancient people living in the Upper Mustang region may have adopted funerary rituals of passing Zoroastrians as they traveled west, Aldenderfer said. These rites, in turn, may have transformed into, or inspired, the Tibetan sky-burial ritual. That idea, according to anthropologist Mark Turin, who wasn't part of the project, is "an interesting and perfectly workable hypothesis."

Covert Caves

The new finds are only the latest to be uncovered in the remote cliffs. In the 1980s, a Nepalese-German team discovered cave tombs dating back about 3,000 years. The human remains in those caves hadn't been de-fleshed, however.

And in 2009, the team behind the new discovery announced they'd found a cliff-cave trove of Tibetan art, manuscripts, and skeletons dating back to the 15th century.

In addition to the newfound mortuary caves, Aldenderfer's team has found nearby caves that were created later, likely for use as living spaces. "I don't think the people who built those 'apartment complexes' actually knew that those mortuary caves were nearby," Aldenderfer said.

SPIRIT OFFERINGS, SOUTHEAST ASIA	PREDATOR BURIAL, EAST AFRICAN MAASAI	SKULL BURIAL, KIRIBATI	CAVE BURIAL, HAWAII
Most people are buried in the fields where they lived and worked. In Cambodia and Thailand, food and drink are left in "spirit houses" for the souls of passed relatives.	While actual burial is reserved for chiefs as a sign of respect, common people are left outdoors for predators.	Skulls of the dead are kept on a shelf in the homes of these traditional islanders, who believe that the native god Nakaa welcomes the dead person's spirit.	A traditional burial occurs in a cave, where the body is formed into a fetal position and covered with a tapa cloth made from mulberry bush bark.

Bloodstained Landscape

Turin, director of the Digital Himalaya Project at the University of Cambridge, said he's not surprised that people have been repeatedly drawn to the Upper Mustang cliffs, despite the challenges.

In fact, the isolation of the cliffs may have been an important part of their allure. Many of the local beliefs that have been practiced in the region, including Buddhism, place great value on the idea of religious retreat, Turin said. "Monks can now practice and reside in monasteries, but we're talking about long before the establishment of any monastery," he said.

"These [caves] may well have been proto-monastic places . . . and as such, people might retreat or bury their dead there."

Also, Turin said, ancient people may have felt tied to the landscape in a way that might be hard for many modern Westerners to understand. Even today, "a well-known story is in circulation about the taming of the territory . . . When the Buddhist saints came up and slew the local deities, their blood and [body parts] stained the Earth and created the colors" of the landscape, Turin said.

> **"The real issue is trying to understand when people came, where they came from, how they got here—all of these kinds of anthropological questions about human migration, movement across landscapes, especially in a landscape like this, which is really difficult."**
> **Mark Aldenderfer**
> *archaeologist*

"The religious culture that exists in people's minds can be mapped onto the landscape. This means that the landscape is sacred, and that caves and places of retreat are similarly sacred." ■

Entrance to Maya Underworld
Found in Mexico?

A labyrinth filled with stone temples and pyramids in 14 caves—some underwater—have been uncovered on Mexico's Yucatán Peninsula.

Stretching south from southern Mexico, through Guatemala, and into northern Belize, the Maya culture had its heyday from about A.D. 250 to 900, when the civilization mysteriously collapsed.

According to Maya myth, the souls of the dead had to follow a dog with night vision on a horrific and watery path and endure myriad challenges before they could rest in the afterlife. The discovery of a subterranean maze has experts wondering whether Maya legend inspired the construction of the underground complex—or vice versa.

Watery Caves

Archaeologists excavating the temples and pyramids in the village of Tahtzibichen, in Mérida, the capital of Yucatán state, have found stones, tall columns, and sculptures of priests in the caves. They've also found human remains.

TRUTH: THE MAYA BELIEVED THAT CAVES WERE THE SOURCE OF THE SUN AND MOON, GODS, AND RACES OF HUMANS, AND WERE THEREFORE CONSIDERED SACRED.

Two team members work at the Maya "underworld" excavation site in Mérida.

In one of the recently found caves, researchers discovered a nearly 300-foot (90-meter) concrete road that ends at a column standing in front of a body of water.

"We have this pattern now of finding temples close to the water—or under the water, in this most recent case," said Guillermo de Anda, lead investigator at the research sites.

"These were probably made as part of a very elaborate ritual," de Anda said. "Everything is related to death, life, and human sacrifice."

Spirit's Quest

Researchers said the ancient legend—described in part in the sacred book *Popul Vuh*—tells of a tortuous journey through oozing blood, bats, and spiders that souls had to make in order to reach Xibalba, the underworld.

"Caves are natural portals to other realms, which could have inspired the Maya myth. They are related to darkness, to fright, and to monsters," de Anda said, adding that this does not contradict the theory that the myth inspired the temples.

> "For the Maya the body was a vehicle for the journey to the afterlife. When a Maya priest made a sacrifice, he was operating in his special universe—helping that universe to continue."
> **Alejandro Terrazas**
> *physical anthropologist*

William Saturno, a Maya expert at Boston University, believes the maze of temples was built after the story.

"I'm sure the myths came first, and the caves reaffirmed the broad time-and-space myths of the Mayans," he said.

Crossing Over and Under

The discovery of the temples underwater indicates the significant effort the Maya put into creating these portals. In addition to plunging deep into the forest to reach the cave openings, Maya builders would have had to hold their breath and dive underwater to build some of the shrines and pyramids.

Other Maya underworld entrances have been discovered in jungles and aboveground caves in northern Guatemala and Belize. "They believed in a reality with many layers," Saturno said of the Maya. "The portal between life and where the dead go was important to them." ■

Tomb of the Otters

Filled With Stone Age Human Bones

Thousands of human bones have been found inside a Stone Age tomb on a northern Scottish island, archaeologists say.

A massive Scottish burial site on South Ronaldsay in the Orkney Islands remained hidden for 5,000 years until a home improvement project accidentally revealed its presence. The ancient grave was found after a homeowner had leveled a mound in his yard to improve his ocean view. Authorities were alerted to the find in 2010 after a subsequent resident, Hamish Mowatt, guessed at the site's significance.

> "It doesn't seem to have been a problem that the otters were living in this tomb at the same time as the community that first built it."
>
> **Dan Lee**
> *Orkney Research Centre for Archaeology (ORCA)*

Mowatt had lowered a camera between the tomb's ceiling of stone slabs and was confronted by a prehistoric skull atop a muddy tangle of bones. "Nobody had known it was an archaeological site before that," said Julie Gibson, county archaeologist for Orkney.

Partial excavation of the site, called Banks Tomb, has confirmed it as the first undisturbed Neolithic burial to be unearthed in Scotland in some 30 years, Gibson reported. "It's certainly unusual to find one whose contents are so well preserved," the archaeologist said. "We have got the assorted remains of many, many people who have been deposited in this tomb at different times."

Entrance to the Tomb of the Otters

Otter Intruders

The underground grave consists of a 4- by 0.75-meter (13- by 2.5-foot) central chamber surrounded by four smaller cells hewn from sandstone bedrock. Capping the central chamber are large water-worn slabs supported by stone walls and pillars.

At least a thousand skeleton parts belonging to a mix of genders and age groups—including babies—have been found to date. Layers of silt divide the remains, suggesting the tomb was in use for many generations, Gibson said.

The site has also been dubbed the Tomb of the Otters because initial excavations revealed prehistoric otter bones and dung amid the human bones. The animal remains indicate that people visited the burial site only sporadically. "It suggests the tomb was not entirely sealed and that otters were trampling in and out a lot" throughout the tomb's use, Gibson said. "For that to occur, you must think there was a gap of a year or two" between grave visits or burials.

Every Bone Tells a Story

So far the excavations, led by the Orkney Research Centre for Archaeology, have "barely scratched the surface," Gibson added. The archaeologist is confident that the site will yield important new clues to Neolithic funerary practices. For instance, researchers hope that DNA and isotopic analysis of the human bones will reveal if the dead were closely related and came from the same tight-knit island community or if the burials include newcomers from overseas.

Archaeologists will also investigate whether bones were removed from the tomb for ritualistic purposes. "This burial is absolutely packed with remains, but with most [other Stone Age tombs], there are actually not that many people in them at the end," Gibson noted.

> # TRUTH:
> ## THE SMALLEST BONE IN THE HUMAN BODY IS SHORTER THAN A GRAIN OF RICE.

Stone Age Not So Aquarian

Meanwhile, recent studies of remains from the nearby Tomb of the Eagles suggest that life among Orkney's Neolithic community of cattle farmers was much less harmonious than previously thought.

At least 20 percent of skulls from that 5,000-year-old site—about a mile from the Tomb of the Otters—show signs of trauma consistent with violent blows from sharp and blunt-edged weapons.

Similar investigations are now being carried out on skeletons from the newfound site. "Neolithic life has had quite a sort of hippy image," Gibson observed. "But it could be that we are looking at ritualized violence." ∎

Headless Romans in England

Came From "Exotic" Locales?

An ancient English cemetery filled with headless skeletons holds proof that the victims lost their heads a long way from home, archaeologists say.

Unearthed between 2004 and 2005 in the northern city of York, 80 skeletons were found in burial grounds used by the Romans throughout the second and third centuries A.D. Almost all the bodies are males, and more than half of them had been decapitated, although many were buried with their detached heads.

York—then called Eboracum—was the Roman Empire's northernmost provincial capital during the time. In a new study of the ancient bones, Gundula Müldner of the University of Reading in the U.K. says the "headless Romans" likely came from as far away as Eastern Europe, and previous evidence of combat scars

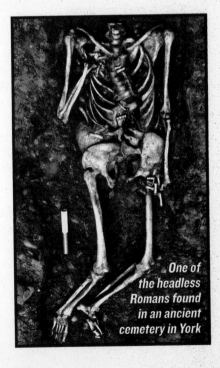

One of the headless Romans found in an ancient cemetery in York

suggests that the men led violent lives. "The headless Romans are very different [physically] than other people from York," Müldner said. "They come from all over the place. Some of them are quite exotic."

They Were What They Ate

Müldner's team analyzed the bones for chemical clues called isotopes, which are different versions of particular elements. Based on the geology and climate of where a person grew up, their bones hold telltale traces of isotopes absorbed from the local food and water.

Oxygen and strontium isotopes in the bones of the headless Romans indicate that just 5 of the 18 individuals tested came from the York area, the team reports in the study, published in the *Journal of Archaeological Science*.

The rest of the men came from elsewhere in England or mainland Europe, possibly from France, Germany, the Balkans, or the Mediterranean.

The suggestion that the headless Romans were a diverse bunch confirms previous archaeological findings. Kurt Hunter-Mann of the York Archaeological Trust, who led the original excavations, said, "We know that the population of Roman York is quite diverse anyway because a lot of traders, for example, were coming from various parts of the Empire," he said. Solving the grisly puzzle of who the headless Romans were will require further bone analysis and forensic studies, still to be completed.

TRUTH: FIFTY-ONE DECAPITATED VIKINGS WERE UNCOVERED IN A THOUSAND-YEAR-OLD EXECUTION PIT NEAR WEYMOUTH, ENGLAND, IN JUNE 2009.

The Millet's Tale

Traces of carbon and nitrogen show that five of the headless Romans ate very different foods from York's local population. And two individuals had a carbon signature from a group of food plants—including sorghum, sugarcane, and maize—not known to have been cultivated in England at that time. "We haven't seen such a signature anywhere in Britain before" in the archaeological record, Müldner said.

In fact, millet is the only food plant from this group that was being grown anywhere in mainland Europe, she added. The archaeologist noted that "the

Romans were not very fond of millet, and often, when they established a new province, other cereals such as wheat would replace millet as the principally grown crop."

Müldner's team thinks the headless millet-eaters hailed from colder climates, perhaps parts of Eastern Europe that were beyond the borders of the Roman Empire. "It might have been the Alps as well, or any higher mountains," Müldner said.

The Soldier Theory

As for what the men were doing in York, previous theories had suggested the headless Romans were slain soldiers, imported gladiators, executed citizens, or ritually killed victims of a religious cult.

Müldner's team favors the military explanation: The ancient city had a large Roman garrison, and the skeletons show injuries consistent with armed combat. It's possible the men were soldiers who had been executed, or who had been killed during battle and had their corpses—with or without heads—recovered for burial by their compatriots.

The Gladiators

Other recent research suggests the headless Romans were gladiators brought to the distant capital for entertainment. Evidence for this notion includes some skeletons' unequal arm development—associated with the specialized use of single-handed weapons—and, on one skeleton, tooth marks from a large carnivore, possibly a gladiatorial lion or bear.

"If the carnivore bite mark is indeed genuine, then, why not, they may indeed have been gladiators," Müldner said. Hunter-Mann says he doubts the new study "will give us conclusive proof one way or another, but it's all very useful." ■

> **"At present our lead theory is that many of these skeletons are Roman gladiators ... One of the most significant items of evidence is a large carnivore bite mark—probably inflicted by a lion, tiger or bear."**
> **Kurt Hunter-Mann**
> *Archaeological Trust*

CHAPTER 2

The Body Human

Nothing may appear as ordinary as your own body, but if you take a closer look, things might start to look a little weird. Your brain, unbeknownst to you, might be "cat napping" while you're awake. Your nose has the ability to sniff out the opposite sex. Your hair can tell scientists if you're a morning person or a night owl. The human body constitutes its own frontier, with scientists discovering strange new things every day.

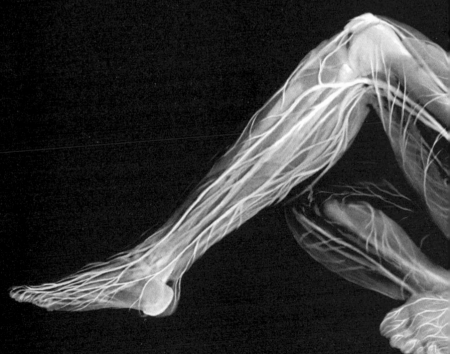

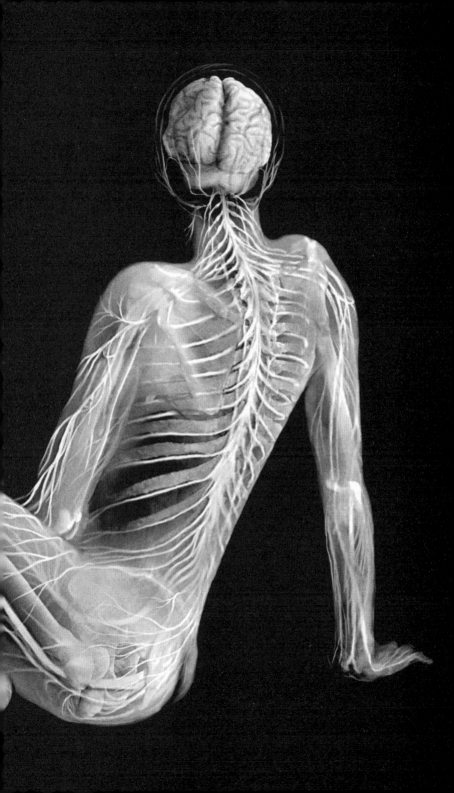

Superhuman Hearing Possible?

Remember *The Bionic Woman* and her incredible eavesdropping abilities? Recent studies show that what was once science fiction could become an everyday reality.

People may one day be able to hear what are now inaudible sounds, scientists say. New experiments suggest that just vibrating the ear bones could create shortcuts for sounds to enter the brain, thus boosting hearing.

What's the Frequency?

Most people can hear sounds in the range of about 20 hertz (Hz) at the low end to about 20 kilohertz (kHz) at the high end. Twenty kHz would sound like a very high-pitched mosquito buzz, and 20 Hz would be what you'd hear if "you were at an R&B concert and you just stood next to the bass," explained Michael Qin, a senior research scientist at the Naval Submarine Medical Research Laboratory in Connecticut. "It would be the thing that's moving your pants leg."

Under certain circumstances, humans can also hear frequencies outside of this normal range. For instance, divers underwater can detect sounds of up to 100 kHz, according to Qin's recent experiments. It's unclear why divers have enhanced hearing underwater, but it may be because the sounds travel directly through the bones to the brain, he said.

> **TRUTH:**
> **DOLPHINS CAN HEAR SOUNDS UNDERWATER FROM 15 MILES AWAY.**

Mechanics of Hearing

In normal hearing, sound waves traveling through the air or water enter our ear canals and strike our eardrums, causing them to vibrate. Our eardrums are connected to three tiny, connected bones called the malleus, incus, and stapes—popularly known as the hammer, anvil, and stirrup, due to their shapes.

As the stapes bone rocks back and forth, it pushes against a fluid-filled structure in the inner ear called the cochlea. Resembling a tiny snail, the cochlea contains tiny hairlike structures that translate the pressure waves in the jostling fluid into nerve signals that are sent to the brain and interpreted as sounds.

Skipping Steps

"If you think of the hearing system as one long chain of events, there are multiple places in which bone conduction or underwater hearing can bypass that chain," he said. For example, bone conduction occurs when very high-frequency sounds directly stimulate the ear bones, sending signals to the brain without activating the eardrums. This is how some species of whales hear underwater.

Hearing Loss Is Widespread

The deaf and hard of hearing account for the single largest group of disabled people in the United States. Of the more than 49 million disabled, at least 28 million have a significant hearing impairment that interferes with communication. That adds up to more than all those with heart disease, cancer, multiple sclerosis, blindness, tuberculosis, venereal disease, and kidney disease combined.

"The core of our work is trying to understand underwater hearing and bone-conduction hearing, and to determine if they share the same underlying mechanism," he said. Alternatively, certain ultrasonic frequencies might stimulate the fluid in the cochlea. "It could be like hitting a wrench against a water tank," Qin explained. "The fluid itself could go into oscillation."

Coming Soon to an Ear Near You?

Qin and his team are now exploring which bones are most likely to be most sensitive to ultrasonic vibrations. He said, "That's the great thing about basic science, right? It lets you know how things work, and you can bend it to many applications."

Could their research lead to devices that give us superhuman hearing or improved hearing aids? Qin is silent, for now. ■

Women Can Sniff Out Men

Without Knowing—and Vice Versa

When it comes time to sort out the men from the women, our noses are giving us information that we don't consciously know about.

Women and men can sniff out the opposite sex via odorless pheromones, a new study suggests. The discovery adds another piece to the growing body of evidence that humans, much like the rest of the animal kingdom, know more from their noses than previously thought.

> **TRUTH:**
> SCENTS SMELL BETTER THROUGH YOUR RIGHT NOSTRIL THAN YOUR LEFT.

Scent of a Woman

Pheromones—chemicals that can communicate sexual information—are widespread in the animal world, and some research suggests humans use them unconsciously as well. "We know that for animals, chemosignals are actually the most used signals to communicate, whereas with humans, we think chemosensation is not really used," said study leader Wen Zhou, a psychologist at the Chinese Academy of Sciences in Beijing. "But based on our experiences, [people] are still influenced by these cues, even if they don't explicitly know it."

In a recent experiment, subjects who smelled possible pheromones from the opposite sex were more likely to interpret ambiguous human figures as that sex—even when the participants didn't know they were smelling anything.

Zhou and colleagues used videos of points of light moving in a way that fools the eye into seeing human motion. The videos were made by filming real people in motion-capture suits with LEDs at each joint—similar to the suits used to create Hollywood special effects. Then the scientists mathematically manipulated the dots until the "figures" had neither a typically male nor typically female gait.

The Very Smell of You

Twenty men and 20 women watched the video animations of these ambiguous figures, as well as ones that were more obviously male or female. While watching the videos, the subjects sniffed clove oil infused with the male steroid androstadienone, the female steroid estratetraenol, or a plain oil used as a base for many cosmetics.

Pheromones Pioneer

Martha McClintock conducted the most prominent study on human pheromones. In 1971 she asserted that pheromones caused women living together to menstruate at the same time, which was a revolutionary idea at the time, because biological processes were not thought to be affected by social factors. McClintock's work has led the way to further research on understanding various human pheromones and the effect of social and psychological contexts on biology.

Men who smelled the female pheromone were more likely to identify the androgynous walker as a woman, and were even more likely to identify more clearly male figures as female than those who just smelled clove oil. The same results applied when women sniffed the male compound: They more frequently saw the ambiguous figures as male than the women who smelled the plain oil. Estratetraenol had no effect on women, and androstadienone didn't affect men.

This perception difference seems to be completely unrelated to what their noses told them: A blindfolded test subject couldn't tell the difference between steroid-infused clove oil and plain oil. "It's completely below their awareness," Zhou said. "They didn't know what they were smelling, but their behavior showed these different patterns." ■

Is Your Brain Asleep

While You're Awake?

If you think you can function on minimal sleep, here's a wake-up call: Parts of your brain may doze off even if you're totally awake.

Sleep deprivation can do strange things to a brain—including putting it on autopilot without your knowledge. In a recent study, scientists observed the electrical activity of brains in rats forced to stay up longer than usual. Problem-solving brain regions fell into a kind of "local sleep"—a condition likely in sleep-deprived humans too, the study authors say.

Surprisingly, when sections of the rats' brains entered these sleeplike states, "you couldn't tell that [the rats were] in any way in a different state of wakefulness," said study co-author Giulio Tononi, a neuroscientist at the University of Wisconsin, Madison.

Despite these periods of local sleep, overall brain activity—and the rats' behaviors—suggested the animals were fully awake. This phenomenon of local sleep is "not just an interesting observation of unknown significance," Tononi said. It "actually affects behavior—you make a mistake." For example, when the scientists had the rats perform a

> # TRUTH:
> THE MAIN SYMPTOM OF FATAL FAMILIAL INSOMNIA (FFI) IS THE INABILITY TO SLEEP, WHICH RESULTS IN EVENTUAL DEATH. ONLY 40 FAMILIES IN THE WORLD ARE KNOWN TO CARRY THE GENE FOR THE DISEASE.

challenging task—using their paws to reach sugar pellets—the sleep-deprived animals had trouble completing it.

Sleep Allows Neurons to Reset?

Tononi and his colleagues recorded the electrical activity of lab rats via electroencephalogram (EEG) sensors connected to the rodents' heads.

As predicted, when the rats were awake, their neurons—nerve cells that collect and transmit signals in the brain—fired frequently and irregularly.

When the animals slept, their neurons fired less often, usually in a regular up-and-down pattern that manifests on the EEG as a "slow wave." Called non-rapid eye movement, this sleep stage accounts for about 80 percent of all sleep in both rats and people.

> "What good does it do to try to educate ... so early in the morning? You can be giving the most ... interesting lectures to sleep-deprived kids early in the morning or right after lunch ... and the overwhelming drive to sleep replaces any chance of alertness, cognition, memory, or understanding."
> **James B. Maas, Ph.D.**
> *sleep expert*

The researchers used toys to distract the rats into staying awake for a few hours—normally "rats take lots of siestas," Tononi noted. The team discovered that neurons in two sections of these overtired rats' cerebral cortexes entered a slow-wave stage that is essentially sleep.

Why Do We Sleep?

It's unknown why parts of an awake brain nod off, though it may have something to do with why mammals sleep—still an open question, said Tononi, whose study appeared in the journal *Nature*.

According to one leading theory, since neurons are constantly "recording" new information, at some point the neurons need to "turn off" in order to reset themselves and prepare to learn again.

"If this hypothesis is correct, that means that at some point [if you're putting off sleep] you're beginning to overwhelm your neurons—you are reaching the limit of how much input they can get." So the neurons "take the rest, even if they shouldn't"—and there's a price to pay in terms of making "stupid" errors, he said.

Even "Alert" People Make Mistakes

Sleep deprivation may have dangerous consequences, Tononi said—and those mistakes may become more common. For one, many people are getting fewer Zs. In 2008, about 29 percent of U.S. adults reported sleeping fewer than seven hours per night, and 50 to 70 million had chronic sleep and wakefulness disorders, according to the U.S. Centers for Disease Control and Prevention. Adults generally need about seven to nine hours of sleep a day, according to the National Sleep Foundation.

What's more, you don't need to feel sleepy to screw up, Tononi emphasized. "Even if you may feel that you're fit and fine and are holding up well," he said, "some parts of your brain may not [be] . . . and those are the ones that make judgments and decisions." ∎

> **TRUTH:**
> STAYING AWAKE FOR 24 HOURS IMPAIRS HAND-TO-EYE COORDINATION AS MUCH AS HAVING A BLOOD ALCOHOL CONTENT OF 0.1%.

Oldest Known Heart Disease

Found in Egyptian Mummy

Heart disease is not a recent development— people's arteries have been clogging up since the days of ancient Egypt.

An ancient Egyptian princess might have been able to postpone her mummification if she had cut the calories and exercised more, medical experts say. Known as Ahmose Meryet Amon, the princess lived some 3,500 years ago and died in her 40s. She was entombed at the Deir el-Bahri royal mortuary temple on the west bank of the Nile, opposite the city of Luxor. The princess's mummified body is among those now housed at the Egyptian Museum in Cairo.

Heart Trouble

Recent scans of 52 of the museum's mummies revealed almost half of the dead have clogged arteries—including the princess. In fact, she is now the earliest known sufferer of coronary athero-sclerosis, a condition caused by a buildup of arterial plaque, which can lead to heart attack or stroke.

Ahmose Meryet Amon— "Child of the Moon, Beloved of Amun"—had blockages in five major arteries, including those that supply blood to the brain and heart, said study co-leader Gregory Thomas, a professor of cardiology at the University of California, Irvine.

"If the princess was in a time machine and I was to see her now, I would tell her to lay off the fat, take plenty of exercise, then schedule her for heart surgery," Thomas said. "She would require a double bypass."

Ancient Heart Attacks

Although the mummies' actual hearts had been removed before entombment, the CT scans uncovered calcium deposits elsewhere in the bodies that are indicative of artery damage. But the study team could not confirm that any of the mummies died of heart disease because most of the mummies' organs were either disintegrated or missing.

Fit for a King

The ancient Egyptians ate a variety of different food, much of which is still consumed today. Some of these foods include hummus, dates, grapes, pomegranates, peaches, watermelon, pork, beef, lamb, goat, duck, and bread. As for beverages, beer was quite commonly served at meals, while the wealthy imbibed wine.

However, a medical text dating back to the time the princess lived—between 1550 and 1580 B.C.—describes the pain in the arm and chest that precedes a potentially fatal heart attack.

In general, blocked arteries and heart attacks are health risks we associate with today's lifestyle and diet, not those of the ancient Egyptians, noted study co-author Michael Miyamoto of the University of California, San Diego's School of Medicine.

"They lacked a lot of the risk factors that we consider to be important in the development of atherosclerosis in modern populations—namely smoking, high rates of diabetes and obesity, and foods rich in trans fats," Miyamoto said.

Too Much of a Good Thing

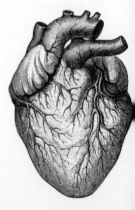

But as a daughter of the Pharaoh Seqenenre Tao II, the princess—like the other examined mummies—was a member of the elite. That means she was possibly more prone to heart disease than the rest of the population.

"Since they were the elite, they presumably led more pampered lifestyles, were more sedentary, and also—maybe importantly—had access to foods which were dense in calories, particularly meats," Miyamoto said.

Adel Allam, study co-author and professor of cardiology at Egypt's Al Azhar University, added that the princess lived during a prosperous period of Egyptian history. "Even the very poor people would eat a lot of pork, and the bread became mixed with honey," he said. "If ordinary people at this time did get a lot of carbohydrates and fat in their diets, then of course the elite would have got even more unhealthy food," he said.

Running in the Family

Perhaps backing up diet as a contributing factor, researchers had previously found some evidence among ancient Egyptians for diabetes, a condition often associated with obesity, Allam added. Also, Egyptian papyrus documents by ancient physicians refer to diabetes symptoms.

However, although body fat is not preserved on ancient mummies, the signs are that Ahmose Meryet Amon was probably petite, Allam said. In fact, the team suspects some factor other than diet and lifestyle may have contributed to her vascular disease.

For instance, "in her family there were a couple of other queens and princesses that had atherosclerosis, so a genetic element cannot be excluded," Allam said. The new study suggests that genetics may be even more important than previously thought in causing atherosclerosis, and the mummies might hold clues to which genetic factors are involved.

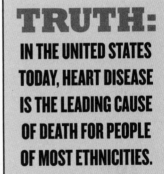

TRUTH:
IN THE UNITED STATES TODAY, HEART DISEASE IS THE LEADING CAUSE OF DEATH FOR PEOPLE OF MOST ETHNICITIES.

Inflammatory Evidence

Another possibility is that atherosclerosis can be brought on by chronic inflammation caused by a person's immune system responding to infection, which in turn can lead to inflammation of the blood vessel walls. "The princess was known to have arthritis and inflammation of the joints," Allam noted. "Also, she had severe dental disease, which is another inflammation."

The U.S.–Egyptian research team will conduct further studies on a total of 72 mummies to investigate the individuals' genetic links to heart issues, as well as evidence of other health issues including arthritis and cancer, and whether their bones can reveal how active they were. ∎

Rejection Really Hurts
Brain Scans Show

Science is showing what we've always known: a broken heart hurts. Brains of the jilted reveal that rejection actually causes physical pain.

Do words ever hurt you as much as sticks and stones? Romantic rejection, at least, causes physical pain, according to a new study of brain activity.

Don't Go Breakin' My Heart

Past studies have shown that simulated social rejection may be connected to a network of brain regions that processes the meaning of pain but not the

Being jilted activates regions of the brain that manage pain.

sensory experience itself. MRI brain scans of people jilted in real life show "activation in brain areas that are actually tied to the feeling of pain," said study co-author Edward Smith, a psychologist at Columbia University in New York City.

Smith and colleagues recruited 40 participants via flyers posted around Manhattan and through Facebook and Craigslist advertisements. All the volunteers reported going through an "unwanted romantic relationship breakup" within the past six months.

While in an MRI machine, the subjects were asked to look at photographs of their ex-partners and think about being rejected. When they did so, the parts of their brains that manage physical pain—the secondary somatosensory cortex and the dorsal posterior insula, to be exact—lighted up, according to the study.

Breaking Up Is Hard to Do

The study isn't a "true perfect experiment—we couldn't control who had the rejection experience and who didn't," Smith noted. "This is true of any study that takes advantage of an activity that happened outside of the laboratory," he said.

"There's always the possibility that there's [some unknown element] about these people who were rejected that was causing the special pattern of what we're seeing."

Yet the results are striking, Smith said, especially because the team analyzed 150 other brain-scan experiments on negative emotions—fear, anxiety, anger, sadness—and found that none of these emotionally painful experiences activate the brain's physical sensory areas in the same way as an undesired breakup. "There may be something special about rejection." ∎

> **"You say that love is nonsense . . . I tell you it is no such thing. For weeks and months it is a steady physical pain, an ache about the heart, never leaving one, by night or by day; a long strain on one's nerves like toothache or rheumatism, not intolerable at any one instant, but exhausting by its steady drain on the strength."**
> **Henry Brooks Adams**
> *American writer*

Cocaine Addiction

Uses Same Brain Paths as Salt Cravings

In your brain, the harmless craving for potato chips or salty pretzels, scientists have found, may share pathways with dangerous addictions to drugs.

Drugs such as heroin and cocaine may owe some of their addictive powers to an ancient instinct—our appetite for salt. In a new study of mouse brains, scientists show that the patterns of gene regulation stimulated by salt cravings are the same gene patterns regulated by drug addiction.

A Million-Year-Old Habit

Salt appetite is a craving millions of years in the making, with likely roots in the salty seas where life on Earth began. "Land dwellers face a problem in that sodium is a trace element, so they have to have a strategy to ingest sodium, and salt craving or sodium appetite is evolution's answer to that," said study co-author Wolfgang Liedtke, an assistant professor of medicine and neurobiology at Duke University.

Salt appetite can be so strong that animals short on sodium will put life and limb at risk to satisfy the hunger. Mountain goats, for instance, are known to cling to sheer cliffs to access a salt lick, even when a misstep means certain death.

The new finding suggests that drug addictions may be so hard to overcome in part because cocaine and opiates—both derived from plants—exploit the brain mechanisms critical for salt appetite.

"Cocaine can usurp the ancient [neural] systems that have made animals better survivors," Liedtke said. The research offers some of the first evidence for addiction processes previously theorized by other experts, added study co-author Derek Denton of the University of Melbourne and the Florey Neuroscience Institute.

Rapid Brain Change

Denton, Liedtke, and colleagues used several techniques (such as withholding salt from test mice or increasing their salt needs by giving them the stress hormone ACTH) to figure out which genes in mammal brains were activated by salt cravings.

The team noticed that, almost as soon as the salt-depleted mice started drinking salt water, the patterns of gene regulation triggered by the need began to reverse. The rapid response is a surprise because it means brain changes in the mice occurred before significant amounts of salt had moved from the stomach to the bloodstream.

> **TRUTH:**
> **PEOPLE HAVE TASTE RECEPTORS IN THEIR LUNGS.**

"It was stunning and perplexing to see that just ten minutes of drinking salty water led to a complete change of the whole sophisticated and elaborate genetic program," Duke's Liedtke said.

Addiction Drivers Still Unknown

It's possible that the research may lead to new treatments for drug addiction that don't rely on "cold turkey" abstinence, which is less likely to be successful against such strong, instinctual cravings.

Overall, though, Liedtke cautions that the new study doesn't fully explain the neural drivers of drug addiction. "Sodium appetite is a healthy instinct. Heroin addiction is a disease that can kill a human," he said. "To go from a healthy instinct to a malady—other things must be happening in the brain." ■

Different Nose Parts for Stinky, Sweet

Tiny hot spots in our noses tell our brains what smells good and what smells yucky.

Millions of receptors in the nose's smelling organ aren't scattered at random. Instead, the receptors congregate in small regions that help the brain discern good smells from bad ones, among other potential functions.

Up Your Nose

The evidence was gathered by sticking electronic probes up people's noses and measuring the chatter of nasal neurons as subjects were exposed to scents.

The findings imply that the pleasantness of a smell is hardwired within our heads, calling into question the impact of life experience on how people perceive smells.

> **TRUTH:**
> HUMANS CAN RECOGNIZE ABOUT 10,000 DIFFERENT SMELLS.

"It's both exciting and disturbing," said Don Wilson of the New York University School of Medicine, a neurobiologist who was not involved in the work. "It doesn't fit in with what I think or a lot of other researchers think about smell." Instead of the brain processing all scent information, for instance, it seems nasal neurons preprocess some of it—almost as if the nose has its own small brain.

Taking One in the Nose

The human nose contains a postage stamp–size smelling organ, called the olfactory epithelium, at the roof of the nasal cavity. By probing mouse noses to measure the firing of nasal neurons, scientists had previously discovered that scent receptors might be organized into groups, similar to the way the tongue has zones armed to detect specific tastes such as sour, sweet, and salty.

But a rodent's nose has more than 1,200 different types of scent receptors, and even a tiny probe could touch tens of thousands of receptors in one reading, making it hard to get a clear signal.

Humans have roughly 400 different kinds of receptors, however, making the business of sticking probes in noses and plucking out useful information more fruitful.

> "Only three receptor types facilitate all of color vision ... For human smell you have 400 [receptors], so it becomes a very complex system to decode."
> **Joel Mainland**
> geneticist, Duke University Medical Center

How the Nose Knows

A team led by neuroscientist Noam Sobel, of the Weizmann Institute of Science in Israel, asked more than 80 people to sniff substances with odors known to be either pleasant or unpleasant across many cultures.

Ordinary smells can be made up of tens to hundreds of compounds, so Sobel and his colleagues puffed only pure chemicals into the noses of their wired subjects, one scent at a time.

Pooling together 801 neural recordings from people's noses, the team found that some regions of the epithelium are better at detecting scent than other regions. The researchers also found hot spots that are better at interpreting either pleasantness or unpleasantness.

"To my surprise, this means something about the epithelium is tuned to pick up certain information" in certain zones, New York University's Wilson said. "We don't understand what the purposes of those zones are," he added, "but that's exciting." ■

Sleep Cherry-Picks Memories

Boosts Cleverness

Your body may be resting during those 40 winks, but your brain is hard at work. The sleeping brain is busy "calculating" what to remember and what to forget, a new study says.

While people are asleep, the brain is selecting what they remember, resulting in sharper and clearer thinking, a new study suggests. Previous research had shown that sleep helps people consolidate their memories, fixing them in the brain so we can retrieve them later. But the new study—a review based on past research on sleep and memory as well as

Alaska Inuit teens sleep in a tent in a file picture.

new studies—suggests that sleep also transforms memories in ways that make them somewhat less accurate but more useful in the long run.

What to Leave In, What to Leave Out

For example, sleep-enabled memories may help people produce insights, draw inferences, and foster abstract thought during waking hours. "The sleeping brain isn't stupid—it doesn't just consolidate everything you put into it, but calculates what to remember and what to forget," said study leader Jessica Payne, a cognitive neuroscientist at the University of Notre Dame in Indiana.

For instance, the memory details that seem to get remembered best are often the most emotional ones, Payne said. Payne and colleagues found that when people are shown a scene with an emotion-laden object in the foreground—such as a wrecked car—they are more likely to remember that object than, say, palm trees in the background, especially if they are tested after a night of slumber.

Rather than preserving scenes in their entirety, the brain apparently restructures scenes to remember only their most emotional and perhaps most important elements while allowing less emotional details to deteriorate.

Don't Sleep on It

Researchers at the University of Massachusetts–Amherst recently conducted a study whose findings suggest that sleeping after a traumatic event may enhance painful emotional memories. "Today, our findings have significance for people with post-traumatic stress disorder, for example, or those asked to give eyewitness testimony in court cases," says Rebecca Spencer, one of the neuroscientists who worked on the study. Interestingly enough, most people find it difficult to sleep after experiencing a troubling event—almost as though the brain does not want to sleep on it.

Picking Memories

Measurements of brain activity support this notion, revealing that brain regions linked with emotion and memory consolidation are periodically more active during sleep than when awake.

"It makes sense to selectively remember emotional information—our ancestors would not want to forget a snake was in a particular location or that someone in the tribe was particularly mean and should be avoided," said Payne, whose study appeared in the journal *Current Directions in Psychological Science.* "Memories are not so much about remembering the past as being able to anticipate and predict multiple possible futures."

> **TRUTH:**
> **YOU'LL SPEND ABOUT SIX YEARS OF YOUR LIFE DREAMING.**

But there are dark sides to such selectivity. For instance, the brain can focus on remembering negative experiences at the exclusion of others, which occurs in depression and post-traumatic stress disorder. Future research may shed light on what details are remembered and how they're remembered, which could help people deal with trauma, Payne noted. "You could also see such work being helpful in coming up with solutions in the classroom or in the business world," she said.

Looking Ahead

Future research may also reveal which components of sleep might be linked with these mental processes. "Does it require the REM sleep associated with dreaming, or deeper slow-wave sleep?" said Robert Stickgold, a cognitive neuroscientist at Harvard Medical School who researches sleep. Overall, "sleep is doing much more complicated stuff than just stabilizing or strengthening memories," Stickgold added.

"We're seeing the sorts of memory processing in sleep that we usually attribute to cleverness." ∎

Drug Could Make Aging Brains More Youthful?

Senior moments and mind fogs might be a thing of the past as a recent study in monkeys hints that declining neural activity can be revved up in older brains.

You can't teach an old brain new tricks—but you can restore its ability to remember the old ones, a new study in monkeys suggests. Chemicals given to rhesus macaques blocked a brain molecule that slows the firing of the brain's nerve cells, or neurons, as we age—prompting those nerve cells to act young again.

"It's our first glimpse of what's going on physiologically that's causing age-related cognitive decline," said study leader Amy Arnsten, a neurobiologist at Yale University. "We all assumed, given there's a lot of architectural changes in aged brains . . . that we were stuck with it," Arnsten said. But with the new results, "the hopeful thing is that the neurochemical environment still makes a big difference, and we might be able to remediate some of these things."

Stress Less!

Intense stress has been shown to have the same effect as aging on the nerve cells in the prefrontal cortex that are responsible for the formation and retention of memories. While the changes caused by stress are reversible, it remains possible that continued exposure to stress increases the decline of memory with age.

The Importance of Working Memory

As the brain gets older, the prefrontal cortex begins to decline quickly. This part of the brain is responsible for many high-order functions, including maintaining working memories—the ability to keep things on a "mental sketch pad" in the absence of stimuli from an action-based task. Researchers previously found that in young brains, nerve cells in the prefrontal cortex excite each other to keep working memories on the brain's slate. "Those connections depend on the neurochemical environment, [which] has to be just right, like Goldilocks," she said.

But when people reach their 40s and 50s, that part of the brain begins to accumulate too much of a signaling molecule called cAMP, which can stop the cells from firing as efficiently—leading to forgetfulness and distractedness. The number of seniors in the United States will likely double by 2050, and many of them will struggle to cope with the frenetic information age, according to the study.

Super Rat

When a lab rat named Hobbie-J was just an embryo, a team of scientists injected her with genetic material that caused an overexpression of the gene NR2B, which helps control the rate at which brain cells communicate. The change allowed Hobbie-J's brain cells to communicate a little bit longer than those of her average counterparts, resulting in higher intelligence. These findings suggest that using drugs to target this gene in humans may help alleviate disorders like dementia and Alzheimer's disease. However, the researchers caution against enhancing memories in healthy people. "There is a reason we forget," says Guosong Liu, a neuroscientist. "We are supposed to leave our bad experiences behind, so they do not haunt us."

Monkey See, Monkey Remember

For their study, Arnsten and her colleagues spent years training six rhesus macaques of various ages how to play simple video games that require the use of working memory. "The youngsters do it great for a long time—they're just like humans," she said. Once the monkeys had mastered the task, the team made recordings of single neurons firing using a tiny fiber inserted painlessly into the brain—a first in any elderly living animal. Not surprisingly, the team found that the younger animals' neurons fired often during periods when there were no stimuli. Neurons in the older animals tended to be less active during the same periods, according to the study, which was published in the journal *Nature*.

But when the team administered certain drugs to the older animals via the fibers—including a chemical called guanfacine—the chemicals blocked the cAMP pathways and revved up neural activity.

Brain Boosters in the Future?

Guanfacine is currently an ingredient in a drug used to treat high-blood pressure in adults. The chemical is also in separate clinical trials to see if it improves working memory in the elderly. Arnsten added that she and her team had led previous studies showing that the drug improved working memory in monkeys, and those results have been repeated by other groups in both monkeys and humans. (It is worth noting that Arnsten does receive royalties from the sale of extended-release guanfacine, called Intuniv, for the treatment of attention deficit hyperactivity disorder in children and adolescents. She does not receive royalties for the generic form of guanfacine being used in the clinical trial.)

Arnsten cautions that even if the drug is approved as a brain booster, it's too early to say how much memory improvement a person could expect. "We can't say it [would] bring you back to being a 30-year-old," she said.

Meanwhile, neuroscientist James L. McGaugh, who was not part of the study team, says that the previous studies "did not, as I understand it,

Drug trials tested the effects on rhesus macaques' working memory.

provide evidence that the enhanced performance was directly associated with 'restored firing' of the neurons. That was an implication."

An Open Question

Importantly, the evidence of enhanced working memory from the previous studies comes from different methods for administering the drugs than those used in the new study, said McGaugh, a fellow at the Center for the Neurobiology of Learning and Memory at the University of California, Irvine.

TRUTH:
BRAIN CELLS LIVE LONGER THAN ANY OF THE OTHER CELLS IN YOUR BODY.

And for the new experiment, in which the drugs were delivered directly to the brain, the team didn't show conclusively whether monkeys' working memory got better after treatment—though past studies have shown a link.

Thus, it's still an "open question" whether more nerve cell activity actually caused memory improvements, McGaugh said. "This is not to question the importance of the findings—just the missing piece of information as I understand their experiments."

Paul Aisen, a neuroscientist and director of the Alzheimer's Disease Cooperative Study at the University of California, San Diego, said the study is "another incremental advance" from a strong group of scientists, but it's "uncertain whether this will have implications for treatment for humans." That's because "measuring . . . firing at the level of a single cell, a neuron, is difficult to extend to human behavior, which is highly complex. It's not so much that a monkey is not a human—it's that this kind of single-cell recording is a very isolated aspect of brain function."

Do We Need Brain Changers?

A bigger question, Aisen added, is whether age-related memory decline really needs drug treatment. "In the absence of a disease such as Alzheimer's, people [compensate] quite well despite the decline in memory," he said. For example, some elderly people combat forgetfulness by simply writing things down.

But study leader Arnsten argues that the fight against cognitive decline is still crucial for many otherwise healthy people. "These abilities are critical for managing one's finances, for being able to manage one's medical treatment, and [to] live independently." ■

Your Hair Reveals if You're a Morning Person

Not sure if you're a morning person or a night owl? Don't worry. Your hair knows.

Early bird or late riser?

The mysteries of your sleep cycle may be unlocked by the hairs on your head, a new study says. That's because the genes that regulate our body clocks can be found in hair-follicle cells, researchers have discovered.

A tiny portion of the brain called the suprachiasmatic nucleus controls the human body clock, and RNA strands—protein-building chains of molecules—process these signals throughout the body in 24-hour cycles.

Predicting Morning People

RNA strands containing the clock genes are found throughout the body—including in white blood cells and inside the mouth—but human hair is easiest for scientists to test.

So Makoto Akashi, of the Research Institute for Time Studies at Yamaguchi University in Japan, and colleagues pulled head and beard hairs from four test subjects at three-hour intervals for a full day.

Time on Your Skin

In addition to hair, our biological clocks can also be detected in our skin cells. Research done at the Institute for Pharmacology and Toxicology at the University of Zurich, Switzerland, found that genes have a major influence on our circadian clock. These cells can provide a precise look at a person's daily body clock by mirroring the molecular operations of the brain's central clock, and such research may lead to new treatments for those who suffer from sleep disorders.

The subjects had already reported their preferred schedules for waking up and eating, among other lifestyle choices. The test day occurred after the subjects had rigorously adhered to their preferred schedules for nine days—in other words, the morning people woke up early every day and the late sleepers woke up late every day.

When the researchers tested the genes in the subjects' follicles, they found that body-clock gene activity peaked right after a subject had woken up, regardless of whether it was 6 a.m. or 10 a.m. This suggests that the brain "turns on" the genes at different times of the morning in different people. Other clock genes followed similar patterns, making it possible to predict "morning people" with just a pluck, the study said.

Body-Clock Disorders

While most people may already know if they prefer to sleep in or wake up early, the new research might also provide insights into human health, researchers say. Disorders of the body clock have been implicated in high blood pressure, diabetes, and even cancer.

The researchers also studied the hairs of rotating shift workers, who are at greater risk for body-clock disorders, for three weeks. During that amount of time, the workers alternated from an early work shift (6 a.m. to 3 p.m.) to a late shift (3 p.m. to midnight).

But the three-week period wasn't enough time for the workers' internal

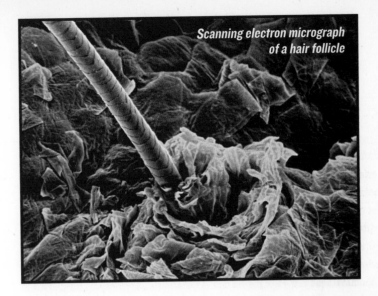

Scanning electron micrograph of a hair follicle

clocks to readjust, according to measurements of follicle genes. Even though the workers' lifestyle was shifted by seven hours, the clock-gene activity in their follicles shifted by only two hours—suggesting shift workers live in a state of jet lag, the study said.

Early Warning Systems?

The follicle test could be used to develop "working conditions that do not disturb clock function" by building in enough time to adjust, the authors wrote. A non-invasive check for a clock disorder could serve as an early warning system. Akashi said: "I hope that our method will be used for regular health checks in schools and companies to keep healthy clocks." ■

TRUTH:
HAIR GROWS ALMOST EVERYWHERE ON YOUR SKIN EXCEPT YOUR LIPS, THE PALMS OF YOUR HANDS, AND THE SOLES OF YOUR FEET.

Mutated DNA

Causes No-Fingerprint Disease

A genetic mutation causes people to be born without fingerprints, experts say.

Almost every person is born with fingerprints, and everyone's are unique. But people with a rare disease known as adermatoglyphia do not have fingerprints from birth. Affecting only four known extended families worldwide, the rare condition is also called immigration-delay disease, since a lack of fingerprints makes it difficult for people to cross international borders.

Coming Up Empty-Handed

In an effort to find the cause of the disease, dermatologist Eli Sprecher sequenced the DNA of 16 members of one family with adermatoglyphia in Switzerland. Seven had normal fingerprints, and the other nine did not. After investigating a number of genes to find evidence of mutation, the researchers came up empty-handed—until a grad student finally found the culprit, a smaller version of a gene called SMARCAD1.

The larger SMARCAD1 is expressed throughout the body, but the smaller form acts only on the skin. Sure enough, the nine family members with no fingerprints had mutations in that gene.

Being born without fingerprints doesn't occur simply because one gene has been turned on or off, Sprecher said. Rather, the mutation causes copies of the SMARCAD1 gene to be unstable.

That mutation is also the first link in a long chain of events that ultimately affects fingerprint development in the womb. The rest of the links in the chain are still a mystery, said Sprecher, of the Tel Aviv Sourasky Medical Center.

No Harm, No Foul

Other inherited diseases that result in a lack of fingerprints—such as Naegeli syndrome and dermatopathia pigmentosa reticularis—are caused by problems with the protein keratin-14.

These conditions "manifest not only with lack of fingerprints, but also with a number of other critical features—a thickening of the skin, problems with nail formation," Sprecher said. By contrast, immigration-delay disease doesn't come with any side effects besides a minor reduction in the ability to sweat. In general, people with the disease "are otherwise completely healthy, like you and me."

TRUTH:
KIDS' FINGERPRINTS DISAPPEAR FROM SURFACES FASTER THAN ADULTS' DO.

Through further study of the Swiss family, Sprecher said, it might be possible to solve the mystery of fingerprints overall. "You go from a rare disease to a biological insight of general importance," he said. "We would never have been able to get to this gene if not for the study of this family." ■

Making Music

Bolsters the Brain's Abilities

Music may be the food of love, but science is showing it nourishes a relationship with your mind as well.

Music may soothe the savage beast, but it can also bring out the best in your brain. "In the past ten years there's been an explosion of research on music and the brain," said Aniruddh Patel, the Esther J. Burnham Senior Fellow at the Neurosciences Institute in San Diego.

Most recently, brain-imaging studies have shown that music activates many diverse parts of the brain, including an overlap where the brain processes music and language. Language is a natural aspect to consider in looking at how music affects the brain, Patel said. Like music, language is "universal, there's a strong learning component, and it carries complex meanings."

Music as Medicine

For decades, music therapy has been practiced to treat a variety of neurological ailments, from Parkinson's and Alzheimer's to anxiety and depression. In addition to improving movement and speech, music can also elicit mood-altering brain chemicals and past memories and emotions.

Above the Din

Do you have trouble hearing people talk at cocktail parties? Try practicing the piano before you leave the house. Musicians—from karaoke singers to professional cellists—are better able to hear targeted sounds in a noisy environment.

Brains of people exposed to even casual musical training have an enhanced ability to generate the brain wave patterns associated with specific sounds,

Playing a musical instrument enhances listening abilities.

be they musical or spoken, said study leader Nina Kraus, director of the Auditory Neuroscience Laboratory at Northwestern University in Illinois.

Kraus's previous research had shown that when a person listens to a sound, the brain wave recorded in response is physically the same as the sound wave itself. In fact "playing" the brain wave produces a nearly identical sound.

But for people without a trained ear for music, the ability to make these patterns decreases as background noise increases, experiments show. Musicians, by contrast, have subconsciously trained their brains to better recognize selective sound patterns, even as background noise goes up.

> "Without music, life would be a mistake."
> **Friedrich Nietzsche**

The overall effect is like a person learning to drive a manual transmission, Kraus said. "When you first learn to drive a car, you have to think about the stick shift, the clutch, all the different parts, but once you know, your body knows how to drive almost automatically."

At the same time, people with certain developmental disorders, such as dyslexia, have a harder time hearing sounds amid the din—a serious problem, for example, for students straining to hear the teacher in a noisy classroom. Musical experience could therefore be a key therapy for children with dyslexia and similar language-related disorders, Kraus said.

The Sound of Music

In a similar vein, Harvard Medical School neuroscientist Gottfried Schlaug has found that stroke patients who have lost the ability to speak can be trained to say hundreds of phrases by singing them first. Schlaug's research demonstrated the results of intensive musical therapy on patients with lesions on the left side of their brain—the area most associated with language.

Before the therapy, these stroke patients responded to questions with largely incoherent sounds and phrases. But after just a few minutes with therapists, who asked them to sing phrases and tap their hands to the rhythm, the patients could sing "Happy Birthday," recite their addresses, and communicate if they were thirsty.

"The underdeveloped systems on the right side of the brain that respond to music became enhanced and changed structures," Schlaug said.

Success varied depending on how recently a person had had a stroke and the severity of the damage, he noted. But several patients were eventually able to teach themselves new words and phrases by turning them into tunes, and a few were even able to move beyond simple phrases and give short speeches. Overall, Schlaug said, the experiments show that "music might be an alternative medium for engaging parts of the brain that are otherwise not engaged."

Northwestern's Kraus agreed. She added that musical training, whatever the age, should be universally encouraged, since it can play a key role in education, clinical therapies, and even in protective measures for keeping the brain sharp as people age. "Plus," she said, "it's just inherently wonderful." ∎

> "Participants report that their control of physical movement improves after playing the drums, their motion becomes more fluid, they don't shake quite as much, and their tremors seem to calm down."
> **Rick Bausman**
> *founder and director,*
> *Drum Workshop, on working with*
> *Parkinson's patients*

The 2,500-Year-Old Brain

How did an unusually well-preserved 2,500-year-old brain stay in such good shape? A team of scientists were determined to solve the mystery.

Scientists may have partially cracked how Britain's oldest known human brain was "pickled" in mud for some 2,500 years. First dug up in 2008 by archaeologists in York, England, the well-preserved chunk of ancient gray matter prompted experts to investigate how the brain had stayed in such good shape.

Brain Preservation

Results of a new study found no chemical signs of deliberate preservation, either by embalming or smoking. Instead, the team suggests that the brain was quickly buried in a pit full of thick, wet clay—among several other factors that may have helped prevent the organ from decomposing.

Protein analysis also confirms the ancient brain—dated to between 673 and 482 B.C.—belonged to a human, said study co-author Matthew Collins, an archaeologist at the University of York. "The majority of the mass of the brain is still there, but it's quite reduced in volume—it's lost a lot of water."

> **TRUTH:**
> IN ANCIENT EGYPT, MUMMIES' BRAINS WERE REMOVED THROUGH THE NOSE.

BRAIN EXCAVATION

1: *York Archaeological Trust workers* **excavate the Iron Age farming site** *where the brain was discovered.*

2: *Rachel Cubitt, of the York Archaeological Trust,* **examines the ancient brain** *using an endoscope.*

3: *One of the pieces of the* **ancient brain** *is seen after removal from the skull.*

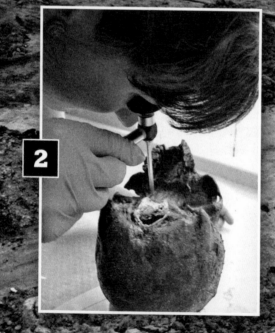

3

Hanged, Then Decapitated

Analyses of the brain and remains of the surrounding skull suggest the Iron Age brain belonged to a male between 26 and 45 who was hanged and then ritually decapitated. The evidence? Gruesome neck-vertebra fractures and cut marks show that hanging occurred before his head was sliced off after death with a sharp instrument, scientists say. The rest of the man's body hasn't been located.

The presumably fatal neck injury "is sometimes referred to as a hangman's fracture, and it is consistent with a long-drop hanging," said study co-author Jo Buckberry, an archaeologist at the University of Bradford. "On the same vertebra there are a series of nine small incisions made by a bladed instrument, which are consistent with a careful decapitation," Buckberry said. "My feeling is the individual would have died from the fracture and the head was removed and deposited in the pit."

> "In the air, even in the chill of a hospital mortuary, brain tissue very quickly decays to liquid before muscle and other soft tissues show much evidence of decay."
> **Sonia O'Connor**
> *postdoctoral research associate, University of Bradford*

A Quick Burial

The fact that the skull was found with intact jaw and neck bones shows the head was buried fresh. "If you moved the skull at a later date, the soft tissue would have decayed"—resulting in the jaw and vertebrae coming detached, said Buckberry, whose study appeared in the *Journal of Archaeological Science*.

In addition to the rapid burial, the cool, low-oxygen conditions of the soil may have aided the brain's preservation, the team found. Cooler temperatures would have slowed the rate at which enzymes degraded the brain tissue, while a lack of oxygen in the soil would have reduced microbial action, the University of York's Collins noted.

Ain't Got No Body

The fact that the head was missing its body may have also kept the brain intact. "Humans that have been buried tend to be eaten from the inside out," Collins said. "Once you've died, the gut bacteria are still pretty hungry."

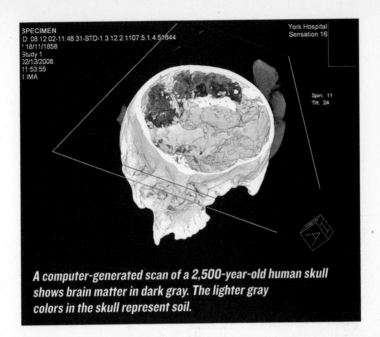

SPECIMEN
D: 08.12.02-11:48:31-STD-1.3.12.2.1107.5.1.4.51644
* 18/11/1858
Study 1
02/12/2008
11:53:55
1 IMA

York Hospital
Sensation 16

Spin: 11
Tilt: 24

A computer-generated scan of a 2,500-year-old human skull shows brain matter in dark gray. The lighter gray colors in the skull represent soil.

These bacteria flood the corpse via blood from the alimentary tract—the digestive pathway through the body—and consume surrounding tissue. "In this case, the blood will have drained from the severed head, and you've no longer got a connection with the alimentary tract."

Still a Mystery

Still, these theories don't fully explain why the brain "didn't turn into mush," Collins said. "It's curious, because normally it is one of those organs that degrade quickly," he said. "There must be something going on internally that we don't understand."

To further investigate the phenomenon, researchers have buried pigs' heads in and around the pre-Roman site, to simulate what may have happened. ∎

Dinosaur, Whole Please

In 1999, then-16-year-old Tyler Lyson discovered the world's most intact dinosaur mummy—the 67-million-year-old hadrosaur, a duck-billed herbivore. In addition to fossilized bones and skin tissue, muscle and organs were also unearthed. It is speculated that the dinosaur was rapidly buried in mineral-rich wet sand, which prevented bacteria from decomposing all of its tissue.

Eating Crocodile
Helped Boost Early Human Brains?

The secret to bigger brains, for early humans anyway, could have been a diet rich in fat, courtesy of crocodiles, a new study suggests.

Eating the fatty flesh of crocodiles may have helped early humans evolve bigger brains, a new study suggests. The work is based on bones and artifacts from a prehistoric "kitchen" that make up the earliest evidence that humans ate aquatic animals.

Brain Food

Stone tools and the butchered bones of turtles, crocodiles, and fish were found at the 1.95-million-year-old site in northern Kenya. No human bones

Fossilized skull of an Euthecodon, a prehistoric crocodile

were found, but the combination of remains suggests early humans used the site specifically to prepare meals.

According to the study authors, the addition of water-based prey into early-human diets may have been what boosted brain size in certain hominins—humans plus human ancestral species and their close evolutionary relatives. That's because reptiles and fish are particularly rich in long-chain polyunsaturated fatty acids. Some experts think this so-called good fat was "part of the package" of human brain evolution, said study leader David Braun, an archaeologist at the University of Cape Town in South Africa.

Discovering evidence for "brain food" in the late Pliocene (about 3 million to 1.8 million years ago) may explain how bigger brains—for instance in our likely direct ancestor *Homo erectus*—arose in humans and their relatives about 1.8 million years ago, Braun said.

Exotic Meats

A number of exotic meats are eaten all over the world and in the United States today, including:

1. Bat (Guam, Indonesia)
2. Kangaroo (Australia)
3. Snake (China)
4. Duck embryo (Vietnam and the Philippines)
5. Crocodile and alligator (Thailand, Australia, parts of the U.S.)
6. Yak (Tibet)

Early Humans Not Crocodile Hunters

Remains of about 48 animal species were found at the Kenyan site, once a delta crisscrossed with small rivers. In addition to the aquatic animals, there's evidence early humans feasted on mammals such as ancient rhinoceroses, hippopotamuses, and antelope, the researchers report in a paper published in the journal *Proceedings of the National Academy of Sciences*. Some of the animal bones bore cut marks from simple, sharp-edged stone tools, according to the study authors.

But the Kenyan hominins were not crocodile hunters, Braun noted. Instead, early humans likely scavenged carcasses, bringing the meat back to the kitchen area to carve up and—in the pre-fire era of human history—eat raw.

Brain Changes

The idea that a diet of aquatic animals "would have been a healthy one in terms of growth and development seems reasonable," Dean Falk, an anthropologist at Florida State University in Tallahassee, said by email. But "the old idea that brain size 'took off' . . . around [two million years ago] has lost support in the last decade," added Falk, who was not involved in Braun's study.

TRUTH:
CROCODILES HAVE BEEN AROUND FOR ABOUT 200 MILLION YEARS.

For instance, a separate study in 2000 led by Falk and published in the *Journal of Human Evolution* found that parts of the brain in some species of the human-ancestor genus *Australopithecus* had started to change shape—a trend associated with an increase in brain size—well before two million years ago. Regardless, study leader Braun said, for the *Homo* genus overall, a diverse menu of mammals and reptiles at some point in human evolution "may be what gave us that adaptive edge." ∎

Astronauts' Fingernails Falling Off

If you're headed for space, you might rethink that manicure: Astronauts with wider hands are more likely to have their fingernails fall off.

A new study reveals that space suit gloves can lead to fingernail troubles—especially for astronauts with wide hands. In fact, fingernail trauma and other hand injuries—no matter your hand size—are collectively the number-one nuisance for spacewalkers, said study co-author Dava

Are space gloves causing the problem?

Newman, a professor of aeronautics and astronautics at the Massachusetts Institute of Technology.

Not Fitting Like a Glove

"The glove in general is just absolutely one of the main engineering challenges," Newman said. "After all, you have almost as many degrees of freedom in your hand as in the rest of your whole body."

TRUTH:
YOUR FINGERNAILS TAKE SIX MONTHS TO GROW FROM BASE TO TIP.

The trouble is that the gloves, like the entire space suit, need to simulate the pressure of Earth's atmosphere in the chilly, airless environment of space. The rigid, balloonlike nature of gas-pressurized gloves makes fine motor control a challenge during extravehicular activities (EVAs), aka spacewalks.

A previous study of astronaut injuries sustained during spacewalks had found that about 47 percent of 352 reported symptoms between 2002 and 2004 were hand related. More than half of these hand injuries were due to fingertips and nails making contact with the hard "thimbles" inside the glove fingertips.

In several cases, sustained pressure on the fingertips during EVAs caused intense pain and led to the astronauts' nails detaching from their nail beds, a condition called fingernail delamination.

While this condition doesn't prevent astronauts from getting their work done, it can become a nuisance if the loose nails are snagged inside the glove. Also, moisture inside the glove can lead to secondary bacterial or yeast infections in the exposed nail beds, the study authors say. If the nail falls off completely, it will eventually grow back, although it might be deformed.

For now, the only solutions are to apply protective dressings, keep nails trimmed short—or do some extreme preventative maintenance. "I have heard of a couple people who've removed their fingernails in advance of an EVA," Newman said.

Pushing Through the Pain

In the current glove design, astronauts wear a pressurized inner layer under a thick outer layer that offers protection from the cold and any passing micrometeorites. On Earth, wearing such space suit gloves might feel like donning a thick set of gardening gloves—a bit restrictive but not too uncomfortable.

"When the glove pressurizes, that nice, flexible fabric surface becomes stiff, like putting air into bicycle tires," said Peter Homer, founder of commercial space suit design company Flagsuit LLC in Maine and two-time winner of NASA's Astronaut Glove Challenge.

"What you find is, depending on the design of the glove, there's pressure on the hard points the hand presses against, and that can give you blisters or cuts," said Homer, who was not involved in the new study. "Also, the materials tend to be rubberized to make the gloves airtight, but that creates a lot of friction against the skin, and that can again create blisters."

During EVAs, astronauts have to work in these gloves for six to eight hours at a stretch. Homer said: "It amazes me that astronauts push through all that pain and get stuff done."

Long Fingers vs. Big Hands

To help design more comfy space suit gloves, MIT's Newman and colleagues initially tested whether fingernail trauma is related to the length of astronauts' fingers. The team first collected data from the Injury Tracking System,

Space Suit Components

How complicated could a space suit be? The answer is very. Officially called the Extravehicular Mobility Unit, or EMU, NASA's space suit is like a personal mini-spacecraft. Here are the different parts that make up an EMU:

1. Primary Life Support Subsystem (PLSS)
2. Upper Torso
3. Hard Upper Torso (HUT)
4. Arms
5. Extravehicular activity (EVA) Gloves
6. Displays and Control Module
7. In-Suit Drink Bag
8. Lower Torso Assembly
9. Helmet
10. Communications Carrier Assembly (CCA)
11. Liquid Cooling and Ventilation Garment
12. Maximum Absorption Garment
13. Simplified Aid for EVA Rescue
14. Wrist Mirror
15. Layers
16. Cuff Checklist
17. Safety Tethers

a database of astronaut medical logs at NASA's Johnson Space Center in Houston, Texas. Of the 232 crew members with complete injury records and body measurements, 22 reported at least one case of fingernail delamination.

Surprisingly, an analysis of hand measurements among injured astronauts and a noninjured control group showed no statistical relationship between finger length and the instances of nails falling off, according to the study.

Instead, the team found that fingernail trauma was a bigger problem for people with a wider hand circumference, as measured around the metacarpophalangeal, or metacarpal, joint, where the fingers meet the palm.

"If you take a pencil and grip it, you're using your metacarpal joint," Newman said. "That's a really difficult thing to repeat when you have a pressurized glove on. A hard palm bar in the soft fabric glove . . . helps make that crease," but the bar also puts pressure on the joint.

The team's analysis, to be published in the journal *Aviation, Space, and Environmental Medicine,* showed that astronauts with hand circumferences greater than about 9 inches (22.8 centimeters)—what Newman called the "large to extra-large range"—had a 19.6 percent chance of fingernail injuries during an EVA. By contrast, astronauts with smaller hand circumferences had just a 5.6 percent chance of losing their nails on the job.

Space Headaches?

A recent survey found that astronauts who did not normally have severe headaches on Earth reported experiencing "exploding" or "heavy-feeling" headaches during spaceflights and extended stays aboard the International Space Station. Although no definitive cause has been determined, there are a few reasons that are likely: "puffy face syndrome," in which fluids that are usually found in the lower extremities shift to other parts of the body; poor air circulation; and perhaps just the same reasons people get them on Earth.

Cold Hands, Hurt Fingernails

"What surprised me is that [the] conventional wisdom [says] that fingernail problems are caused by repetitive tapping on the fingernail . . . and you'd think that if you had longer fingers, you'd be banging on the end of the glove more," said Flagsuit's Homer.

But the hand-width hypothesis "is good, too," he said. "The bigger the hand is, the more the glove squeezes on [the metacarpophalangeal] joint and cuts off blood flow."

Damage to the tissue underneath the fingernail could also occur if circulation at the knuckle joint is repeatedly shut off and then restored, which could also lead to delamination. This could also explain why so many astronauts have reported that their fingertips get cold during EVAs despite their thermal gloves, Homer said. Overall, he added, the new paper "shines light on a whole new direction on how to address this issue." According to Homer, the key is to make all parts of a glove custom fitted for each astronaut.

A Custom Fit

For anyone selected for an EVA, the airtight inner layer for the current glove design is custom made via hand casts, laser scanning, computer modeling, and special machining techniques. But the outer layer is built in discrete sizes—more like a "small, medium, large" situation, he said.

"It costs around a hundred thousand dollars up front to custom fit the airtight bladder," Homer said. "In my opinion that also needs to apply to the outer layer, which really gives the glove its shape."

Customization may not always solve the issue, though, MIT's Newman said: "Some may like a tighter or looser fit—there's variability in subjective desires. And if you have a really tight fit, you're going to have a lot more pressure" on the metacarpal joint.

Robots and Shrink Wrap?

Newman thinks another option worth looking into is robotic amplification inside the glove. "Say I'm grabbing on to something. I'm using muscles to act against gas-pressurized gloves," she said. "But what if I had little actuators in there? My fingers can do less work—that'd be great!"

She added that "there are design trade-offs" to robotic gloves. "But we have some big dreams here: small-mass systems close to the skin that work in concert with muscles and bones, not big clunky exoskeletons."

Newman and colleagues have also been experimenting with entire skin-tight space suits that rely on mechanical counterpressure: Rather than working in a gas-pressurized bubble, astronauts would effectively get shrink-wrapped in a suit made of flexible material.

No matter what the approach, Newman said, "the bottom line is we want people to be working in a space suit glove that's working with them, not against them." ■

> **TRUTH:** ASTRONAUTS GROW UP TO THREE INCHES TALLER IN OUTER SPACE.

Ball Lightning

May Be a Hallucination

Mysterious floating blobs of light known as ball lightning might simply be hallucinations caused by overstimulated brains, a new study suggests.

For hundreds of years eyewitnesses have reported brief encounters with golf ball– to tennis ball–size orbs of electricity. But scientists have been unable to agree on how and why ball lightning forms, since the phenomenon is rare and very short-lived.

Ball Lightning Basics

Eyewitnesses have described ball lightning as a floating, glowing ball similar in size to a tennis ball or even a beach ball. The sightings generally accompany thunderstorms, but it's unclear what other similarities ball lightning might share with its conventional relative.

Ball lightning floats near the ground, sometimes bounces off the ground or other objects, and does not obey the whims of wind or the laws of gravity. An average ball lightning glows with the power of a 100-watt bulb. Some have been reported to melt through glass windows and burn through screens. The phenomenon lasts only a short time, perhaps ten seconds, before either fading away or violently dissipating with a small explosion.

> **TRUTH:**
> THERE ARE ABOUT 3,000 LIGHTNING FLASHES ON EARTH EVERY MINUTE.

The record suggests that ball lightning is not inherently deadly, but there are reports of people being killed by contact—most notably the pioneering electricity researcher Georg Richmann, who died in 1753. Richmann is believed to have been electrocuted by ball lightning as he conducted a lightning-rod experiment in St. Petersburg, Russia.

In the Mind's Eye?

Because ball lightning is often reported during thunderstorms and it's known that multiple consecutive lightning strikes can create strong magnetic fields, Joseph Peer and Alexander Kendl at the University of Innsbruck in Austria wondered whether ball lightning is really a hallucination induced by magnetic stimulation of the brain's visual cortex or the eye's retina.

In previous experiments, other scientists had exposed humans to strong, rapidly changing magnetic fields using a medical machine called a transcranial magnetic stimulator, or TMS. The machine's magnetic fields are powerful enough to induce electric currents in human brain cells without being harmful.

Focusing magnetic fields on the visual cortex of the brain caused the subjects to see luminous discs and lines. When the focus was moved around within the visual cortex, the subjects reported seeing the lights move.

In their paper, which appeared on the physics research website arXiv. org, Peer and Kendl argue that magnetic fields made by lightning could

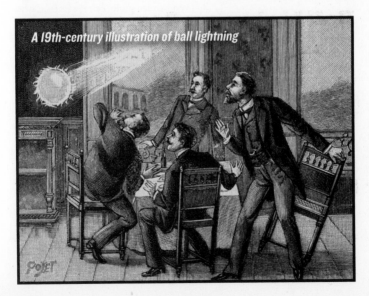

A 19th-century illustration of ball lightning

have the same effect as TMS machines on nearby humans. In fact, the pair thinks about half of all ball lightning reports are actually tricks of the mind induced by magnetism.

Not All Imaginary

The researchers make a convincing argument that some ball lightning reports are spurred by hallucinations, said John Abrahamson, a chemist and ball lightning expert at the University of Canterbury in New Zealand who was not involved in the study.

But "I cannot believe that most of the images reported as ball lightning are due to this brain influence," Abrahamson said in an email. For one thing, the colors of light seen by the subjects in the experiment were "white, gray, or in unsaturated colors." But ball lightning has been reported in a variety of colors, including orange, green, and blue, Abrahamson said.

> **TRUTH:**
> **THE AIR AROUND A LIGHTNING STRIKE IS FIVE TIMES HOTTER THAN THE SUN.**

Also, some eyewitness reports of ball lightning include close-up observations detailing the internal structures of the balls and even associated smells and sounds. Some reports of ball lightning even involve multiple eyewitnesses who saw the same phenomenon from different angles and saw the balls travel in the same directions.

"This common geometric perception from different angles would be very unlikely if their brains were being stimulated" by the local magnetic field caused by lightning strikes, Abrahamson said.

Lighting Up the Lab

Eli Jerby, an engineer at Tel Aviv University in Israel, has actually created something similar to ball lightning in the lab. He also doesn't think hallucinations could account for all ball lightning reports. "While hallucinations could explain some cases, the effect of ball lightning is yet feasible in both nature and the laboratory," Jerby said in an email.

"Furthermore, with the recent experimental progress by us and by others, we are closer than ever to simulating natural ball lightning completely in the lab, and to explaining the real ball lightning enigma" in nature. ■

Secrets of Synesthesia

Neural tangling called synesthesia may have creative benefits, experts say.

A neural condition that tangles the senses so that people hear colors and taste words could yield important clues to understanding how the brain is organized, according to a new review study.

This sensory merger, called synesthesia, was first scientifically documented in 1812 but was widely misunderstood for much of its history, with many experts thinking the condition was a form of mild insanity.

The Color of Two

"It's not just that the number two is blue, but two is also a male number that wears a hat and is in love with the number seven," said study co-author David Brang, of the University of California, San Diego (UCSD). "We're not sure if these personifications are [also a symptom of] synesthesia, but we think this is what derailed a lot of scientists from being interested in it . . . They thought these people were making it all up."

Early misunderstandings of synesthesia were due in part because the associations that synesthetes

> "The taste of beef, such as a steak, produces a rich blue . . . Mango sherbet appears as a wall of lime green with thin wavy strips of cherry red. Steamed gingered squid produces a large glob of bright orange foam, about four feet away, directly in front of me."
>
> **Sean Day**
> *linguistics professor, National Central University, Taiwan*

described were very precise and detailed, prompting some experts at the time to link the condition with mental disorders such as schizophrenia. Another early "view held that synesthesia was a 'throwback' to a more evolutionarily primitive state," said study co-author Vilayanur Ramachandran, also a neuroscientist at UCSD.

During the past 30 years, though, a growing body of evidence has shown that synesthesia has a physical basis—for example, the brains of synesthetes are wired differently, and the condition is highly heritable, which indicates there is a genetic component.

In fact, the study authors think it's possible such a strange phenomenon has survived in an evolutionary sense because it offers people certain benefits to creative thinking. "Ninety-five to ninety-nine percent of synesthetes love their synesthesia and say it enhances their lives," Brang said.

A Better Understanding

Today scientists have tools that allow them to probe the brain in ways that were impossible 200—or even 10—years ago. One such tool is a type of brain scan called DTI, short for diffusion tensor imaging, which lets scientists see the connections between different brain regions. "We can see that there's increased connections in synesthetes between the associated [senses]," Brang said.

Visualizing these connections between sensory brain regions could help explain why certain forms of synesthesia exist and why the condition tends to be unidirectional—for example, numbers can evoke colors but colors don't typically evoke numbers. Such studies could also help test an idea proposed by some scientists that all humans have the neural mechanism for synesthesia but it's suppressed for some reason.

Another positive development in synesthesia research is that scientists are relearning how to listen to their subjects, the study authors say.

"Listening to the subjective reports of people fell out of practice in the mid-20th century, but you can learn an amazing amount of information

> **TRUTH:**
> **IN THE UNITED STATES, STUDIES SHOW THAT THREE TIMES AS MANY WOMEN AS MEN HAVE SYNESTHESIA; IN THE UNITED KINGDOM, EIGHT TIMES AS MANY WOMEN ARE REPORTED TO HAVE IT.**

from just sitting down for 20 minutes and talking to a patient," Brang said. "You can begin to trust what they're experiencing."

A Boon to Creativity?

Studies today indicate that synesthesia is about seven times more common in artists, poets, and novelists than in the rest of the population, and some scientists have hypothesized that synesthetes are better at linking unrelated ideas.

"We worked with a novelist years ago who swore that her synesthesia helped her pick metaphors," Brang said. "She said she would know what color a word should be even before she knew what the word was."

Some savants with synesthesia have been known to perform amazing feats of memorization, such as remembering the value of pi to 22,514 digits. Other synesthetes are able to distinguish between very similar colors or have a heightened sense of touch.

Despite recent advances, many questions about synesthesia remain, such as whether other animals experience synesthesia, how different brain chemicals affect the condition, and the exact role of genetics in determining a synesthete's cognitive and creative abilities. Also, Brang said, "if it's so cool and such a great trait, why don't we all have it?" ■

> ### Famous Synesthetes
>
> 1. Vasily Kandinsky (painter, 1866–1944)
> 2. Olivier Messiaen (composer, 1908–1992)
> 3. Charles Baudelaire (poet, 1821–1867)
> 4. Franz Liszt (composer, 1811–1886)
> 5. Arthur Rimbaud (poet, 1854–1891)
> 6. Richard Phillips Feynman (physicist, 1918–1988)

Secrets of Sleeping Soundly

Sleep like a log? You can thank your "lucky spindles," your brain's blockades against slumber-disrupting noises.

While some toss and turn, others can sleep like the dead. The difference may lie in their brains—specifically in their spindles, rapid-fire brain waves that act as blockades against noise during sleep, a new study says.

Sleep Science

For the research, study co-author Jeffrey Ellenbogen of Harvard Medical School recruited 12 self-described sound sleepers to spend three nights in his "comfy" lab.

On the first night, the sleepers were treated to quiet, idyllic conditions.

But during the next two nights, scientists bombarded the subjects with several types of loud sounds—including jet engine roars and toilet flushes—after the people had fallen asleep. Brain wave readings revealed that the more spindles a person had, the more likely he or she could stay asleep through the barrage of noises, Ellenbogen said.

Everybody has spindles, which are controlled by the thalamus, a "way station" that conveys sensory information to other parts of the brain, Ellenbogen said. But there's still much that's unknown about these "sleeping" brain waves. For instance, it's a mystery why some people have more spindles than others, Ellenbogen said.

TRUTH:
THE LONGEST A PERSON HAS GONE WITHOUT SLEEP IS TEN DAYS.

Certain parts of the brain control how soundly you sleep.

Seeking a "Sleep Utopia"

This new research may bring Ellenbogen and colleagues closer to creating a "sleep utopia" for troubled slumberers, he noted. Fractured sleep—when a person wakes up many times a night—is "disturbingly prevalent in our society, partly due to insults from a variety of noises," according to the study.

In addition to early-morning garbage trucks and creaking pipes, people are increasingly surrounded by technology that may produce irritating "beep and boops," Ellenbogen said. "Now we can leverage this naturally occurring process [of spindle generation] and use that as a tool to prevent the sleeper from disruption," he said. For instance, it may be possible to design a drug that would enhance spindles in light sleepers.

In the meantime, testing a person's spindle activity may help predict an individual's tolerance to noise, Ellenbogen added. This could help with life decisions, he said, such as, "Should I take the job that puts me in the city, where I'm [in] urban chaos?" ■

CHAPTER 3

Crea
Features

ture

Whether cuddly or creepy, furry or fanged, animals are odd. Some creatures just look weird—like the recently discovered tube-nosed fruit bat (nicknamed "Yoda") from Papua New Guinea. Others have spooky super-powers—like vampire bats with heat-seeking vein sensors. Others engage in strange activities, like the Italian goats that like to climb up the sheer vertical faces of dams. Every day, animals are revealing to us new ways to be weird, each one more surprising than the next.

Five Weirdest New Animals

There are new things under the sun, including some of the strangest animals we've ever seen.

Whether deep in the world's jungle or in the restaurants of Vietnam, scientists and explorers are still discovering the most unusual creatures in the world in the most unexpected places.

ANIMAL 1 — "Yoda Bat"

This tube-nosed fruit bat—which became a web sensation as "Yoda bat"—is just one of the roughly 200 species encountered during two scientific expeditions to Papua New Guinea in 2009, scientists announced.

Though seen on previous expeditions, the bat has yet to be formally

Yet to be formally documented as a new species, this fruit bat is.

documented as a new species, or even named. Like other fruit bats, though, it disperses seeds from the fruit in its diet, perhaps making the flying mammal crucial to its tropical rain forest ecosystem.

ANIMAL 2 — *The Simpsons* Toad

Nosing around for "lost" amphibian species in western Colombia in 2010, scientists stumbled across three entirely new species—including this beaked toad. "Its long, pointy, snoutlike nose reminds me of the nefarious villain Mr. Burns from *The Simpsons* television series," said expedition leader Robin Moore.

The unnamed, 0.7-inch-long (2-centimeter-long) toad is "easily one of the strangest amphibians I have ever seen," added Moore, an amphibian-conservation specialist for Conservation International. The toad also has an odd reproductive habit: skipping the tadpole stage. Females lay eggs on the rain forest floor, which hatch into fully formed toadlets.

This "excellent" little toad is said to resemble Mr. Burns from The Simpsons.

ANIMAL 3 — Self-Cloning Lizard

You could call it the surprise du jour: A popular food on Vietnamese menus has turned out to be a lizard previously unknown to science, scientists said in 2010.

What's more, the newfound *Leiolepis ngovantrii* is no run-of-the-mill reptile—the all-female species reproduces via cloning, without the need for male lizards.

About 1 percent of lizards can reproduce by parthenogenesis, meaning the females spontaneously ovulate and clone themselves to produce offspring with the same genetic blueprint.

The newfound lizard is a common food in southeastern Vietnam.

"The Vietnamese have been eating these for time on end," said herpetologist L. Lee Grismer of La Sierra University in Riverside, California, who helped identify the animal. "In this part of the Mekong Delta [in southeastern Vietnam], restaurants have been serving this undescribed species, and we just stumbled across it."

ANIMAL 4 "Demon" Bat

Meet a new prince of the underworld—the Beelzebub bat. Named for its diabolic coloration, the recently discovered

Beelzebub bats are shy and avoid contact with humans.

bat has a black head and dark back fur, both of which contrast sharply with the flyer's whitish belly, scientists reported in a 2011 study.

Despite the fiendish name, Beelzebub bats are typically shy creatures, doing their best to avoid humans in their remote rain forest habitat in Vietnam, scientists say. If captured, however, the bats can turn fierce, said study co-author Neil Furey, a biologist with the conservation group Fauna & Flora International.

ANIMAL 5 | Hispaniolan and Cuban Solenodons

This may look like a rodent of unusual size, but the rare Hispaniolan solenodon isn't a rodent at all. More closely related to shrews and moles, the solenodons are the only mammals that inject prey with venom, through special grooves in their teeth. There are only two species: the Hispaniolan solenodon—native to the island of Hispaniola, shared by Haiti and the Dominican Republic—and the Cuban solenodon, which was rediscovered in 2003.

Until the introduction of predators such as dogs, cats, and mongooses to their island habitats, the "slow and clumsy" solenodons had no natural enemies, according to EDGE of Existence, a global conservation initiative. "Once in the hand, they will do their best to escape," he said. "In essence, they exhibit a 'flight' first and 'fight' second response—the latter only when they have no other option." ■

Venomous mammals, like the solenodon, are a rarity.

Crocodile Attacks Unsuspecting Elephant

It was an African ambush that became an Internet sensation. A mother and child elephant at a water hole had a deadly encounter with a crocodile. Who came out on top?

TRUTH:
CROCODILES CAN'T CHEW.

A routine trip to the water hole resulted in a life-or-death struggle for a pair of African elephants when they were ambushed by a hungry Nile crocodile. As the two approached the water, the croc launched out of the water and seized the adult's trunk. Tourist Martin Nyfeler of Kloten, Switzerland, captured pictures of the wild encounter during a visit to Zambia's South Luangwa National Park.

"We saw a mother elephant and baby at the water hole and said [to the guides], 'You know, what a cute picture, let's stop here,'" Nyfeler told National Geographic News. "And suddenly the croc jumped out. The whole event took maybe 15 seconds."

Size Matters?

Although elephants are very unusual prey for Nile crocodiles, the 20-foot-long (6-meter-long) reptiles will occasionally ambush and take down large animals—including dozens of people annually, experts say.

"Even as [crocodiles] get bigger, most of their diet will be fish or smaller animals," said Jason Bell, assistant curator of reptiles and amphibians at the Philadelphia Zoo.

"But they are also an opportunistic predator that will wait for something to come to the water's edge and drink," he said. "They've been known to take down young hippos or Cape buffalo—that's a huge animal that they can pull into the water."

Drag Out Fight

During the attack, the elephants were able to move quickly away from the water with the crocodile still hanging on to the adult. According to photographer Nyfeler, guides in the Zambian park had never before seen such an encounter.

Even for the formidable Nile crocodile, bringing down an elephant is no easy task—suggesting the ambush may have been either an act of desperation or perhaps a simple miscalculation, according to Don Boyer, San Diego Zoo's curator of herpetology.

"Predators can make mistakes," Boyer said. "They can take on something and then say, Wow, hindsight is 20/20, and this was a big mistake."

"I think there is a misconception about elephants. And the incidents of elephant rage come from elephants in a disturbed population or in conflict with people. I've found that if you give them the benefit of the doubt, they are not generally an aggressive species. They are naturally gentle and trusting. When you betray that trust, they get aggressive."
Martyn Colbeck
filmmaker, on shooting with elephants

Elephant Escape

This particular clash of the titans had a happy ending—except perhaps for the hungry crocodile. "The elephant managed to turn, but the croc was still hanging on," photographer Nyfeler said. "Then the little baby somehow stumbled over the croc, and the croc released the elephant. "The croc went back into the water, and both elephants just ran away."

ELEPHANT ESCAPE

1: *A tourist captured the crocodile's surprise attack on a pair of elephants.*

2: *A vicious tug-of-war ensues between crocodile and elephant.*

3: *The baby elephant stumbles over the croc, causing it to release its grip.* ■

1

2

3

Spiky Rat Plant Poison
Turns Hair Deadly

An East African murder mystery has a solution: It was the crested rat, in East Africa, with the poison bark.

You've heard of rat poison, but poisoned rats? In East Africa, a porcupine-like rat turns its quills into lethal weapons by coating them with a plant toxin, a new study says. Neighboring African hunters use the same substance to make elephant-grade poison arrows. No other animals are known to use a truly deadly external poison, researchers say.

TRUTH:
A CHEMICAL RELATED TO OUABAIN, CALLED DIGITOXIN, HAS BEEN USED FOR DECADES AS A TREATMENT FOR HEART FAILURE.

A Rat's Arsenal

Scientists have long suspected that the crested rat might be using poison because of stories of dogs becoming ill or dying after encounters with the rodent, and because it has a distinct black-and-white warning coloration seen in other species.

It was unclear until now, however, where the nocturnal rat got its poison. The researchers made their discovery after presenting a wild-caught crested rat with branches and roots of the *Acokanthera* tree, whose bark includes the toxin ouabain.

The animal gnawed and chewed the tree's bark but avoided the non-toxic leaves and fruit. The rat then applied the pasty, deadly drool to spiky flank hairs. Microscopes later revealed that the hairs are actually hollow quills that rapidly absorb the ouabain-saliva mixture, offering an unpleasant surprise to predators attempting to taste the rat.

Poison isn't the only item in the armory of the roughly foot-and-a-half-long (45-centimeter-long) rat, which lives in burrows in East Africa, the team says. Very tough skin and a reinforced skull that looks like "it could take a couple of head shots" also suggest the rodent doesn't shrink from a fight, said study co-author Tim O'Brien, a biologist with the Wildlife Conservation Society. And with leopards, honey badgers, jackals, and wild dogs among its assumed predators, the rat may need all the advantages it can muster.

5 Animals Deadliest to Humans

1. Mosquito: causes 2 million deaths annually
2. Asian cobra: causes majority of snakebite deaths in India
3. Australian box jellyfish: kills with one of world's deadliest toxins
4. Australian saltwater crocodile: believed to cause 2,000 deaths per year
5. Hippos: kill more people in Africa than any other mammal

The crested rat, native to East Africa, uses poison bark for defense.

Paralyzing Poison

"For as long as colonials have been in Kenya, there have been tribes that specialize in hunting elephants using this poison, and now we're seeing it being used in rhino poaching," O'Brien said. Ouabain can cause paralysis and, at high enough doses, heart attacks.

"The dogs that we know of that have attacked the crested rat have suffered everything from temporary paralysis for a couple of weeks to death," said O'Brien, who is also a researcher at the Mpala Research Centre in Kenya, where the experiment was conducted.

Other mammals are known to use toxins that they don't produce themselves. For example, a hedgehog species applies a mild toxin from a toad to its fur, O'Brien said. Likewise, some capuchin monkeys rub an extract from millipedes onto their fur to repel insects. "But there are no other examples of an animal using [an external substance] that is actually a lethal toxin," he said.

> "What is quite clear in this animal is that it is hardwired to find the poison, it is hardwired to chew it and it is hardwired to apply it to the small area of hairs."
> **Jonathan Kingdon**
> *researcher at Oxford University*

Why Is the Rat Immune?

While the new study solves one crested rat mystery, others remain. It's unclear, for example, why the rodent itself doesn't suffer from the poison. One idea is that its saliva subtly alters the poison, O'Brien said.

Scientists also don't know how each young crested rat learns about ouabain. Is the chemistry lesson passed from parents to offspring? Does each animal discover the poison on its own? "No one knows," he said. "We know the general outlines about this animal but none of the details." ■

Vampire Bats

Have Vein Sensors

Here's a finding that might make your blood run cold—vampire bats have specially evolved nerves that can sense the heat of your veins.

The facial nerves of vampire bats tell them much more than we ever thought they could: They tell them where the veins are. Scientists already knew that vampire bats have snakelike pit organs in their faces that point the mammals to the juiciest parts of their prey—the veins. But it was unknown how the predators located those choice biting spots, until now.

The vampire bat's facial nerves tell it the best places to bite.

Bloodsucker's Senses

Now a study has shown that bats have evolved special facial nerves that can detect body temperatures as low as 89.6 degrees Fahrenheit (32 degrees Celsius). The vampire bat is "clearly adapted in a lot of unusual ways for a very unusual lifestyle—this is one more example," said study co-author Nicholas Ingolia, a genomics researcher at the Carnegie Institution for Science in Baltimore, Maryland.

People have a similar heat-sensitive channel, but it's only activated by painfully hot stimuli, such as touching a hot stove. The vampire bat, the study found, has two forms of heat-sensitive channels, one to sense painful heat—like us—and another to zero in on its prey's veins.

Blood Thinner

An enzyme found in vampire bats' saliva, desmoteplase, or DSPA, is being used in a new experimental drug treatment for stroke victims. DSPA helps thin blood and prevent clotting—a useful tool for vampire bats to get the most blood out of each bite. The current medication for dissolving blood clots that cause brain attacks, TPA (tissue plasminogen activator) must be used within three hours after the onset of symptoms, so most stroke victims don't receive the treatment in time. On the other hand, DSPA can be used up to nine hours after the onset of symptoms and is easier to administer.

Bloody Nerve Cells

Such an evolutionary adaptation—known previously only in three snake species—is especially crucial for the vampire bat, which needs a blood meal every one or two days to stay alive. The vein-sensing ability is "an extreme version of an existing trait," Ingolia said. Other bats have the gene for this hypersensitive channel, but only the bloodsuckers' bodies appear to activate the gene.

In the study, Ingolia and colleagues isolated the nerve cells that travel to the pit organs in the bat's face and compared them with the sensory nerve cells that go to the rest of the bat's body. The nerve cells were anatomically different from the regular, pain-sensing nerve cells, which means the vein-sensing cells convey separate information, Ingolia said. The team also looked at vampire bat genes and found that the same heat-sensing gene is found throughout the body, but the gene results in a different type of nerve cell in the pit organs. ∎

Goats Scale Dam Walls

Salt cravings drive goats to new heights—literally.

Using moves that would make any rock climber jealous, Alpine ibex scaled a near-vertical rock face of a northern Italian dam. In summer 2010, pictures of the goats were making the rounds online, particularly in emails falsely claiming the animals are bighorn sheep on Wyoming's Buffalo Bill Dam (the rumor-quashing website snopes.com busted the hoax).

Crazy Salt Cravings

In truth, Adriano Migliorati snapped the pictures at the 160-foot-tall (49-meter-tall) Cingino Dam, the Italian hiker told National Geographic News via email. The goats are attracted to the dam's salt-crusted stones, according to the U.K.-based Caters news agency. Grazing animals don't get enough of the mineral in their vegetarian diets and scale the dam to get a fix.

It's not far-fetched, though, to think such a scene could be photographed in the United States. For example, mountain goats could scale dams in the U.S. West, according to Jeff Opperman,

TRUTH:
GOATS' EYES HAVE RECTANGULAR PUPILS.

senior adviser for sustainable hydropower at the U.S.-based nonprofit the Nature Conservancy.

Opperman, who called the Cingino pictures "mind-boggling," described a similar instance of a Montana mountain goat doing an "incredibly acrobatic

To satisfy their salt cravings, several goats scale Cingino Dam in Italy.

stretching maneuver to lick salt . . . He is wedged up this sheer vertical cliff face, almost doing a yoga pose with four hooves splayed out there," he said. "It's the same concept [with the Italian goats]—these animals can overcome what looks like impossible topography to get what they want."

Opperman cautioned, though, that the Italian dam is rare, in that its rough masonry provides gaps that act as toeholds. The more common, smooth-concrete dams—such as Buffalo Bill Dam—would give goats anywhere in the world trouble, he said.

Gravity-Defying Goats

Cingino Dam isn't completely vertical, allowing ibex to gain some purchase. Adapted to their perilous environment, Alpine ibex have evolved a specialized split hoof, whose cleft is wider than on any other split-hooved species, according to *Smithsonian* magazine. The hoof also has a hard wall that can grab on to steep cliffs and a soft, rubbery inside that serves as a "stopper" when the animal is pushed forward by gravity, the magazine reported.

And because dams are usually built in steep canyons, Cingino's steep rock face is likely nothing novel for the mountain-dwelling ibex, according to Opperman. The herbivores spend their lives scrambling the European Alps' rocky and steep terrain.

Close Call

The alpine ibex nearly became extinct in the early 19th century; it was believed to have healing powers, which led to overhunting. Today, the ibex is no longer endangered as a result of conservation and reintroduction of the species in Switzerland, Austria, and France.

By scaling Cingino Dam, salt-craving ibex are "showing ingenuity, taking advantage of this human-created thing in their environment," the Nature Conservancy's Opperman said. ■

Nature's Four Worst Mothers

Who are the worst moms in the animal kingdom? Do they deserve our judgment, or should we cut them a break?

There are a number of outstanding moms in the animal kingdom, but many animal mothers are more *Mommie Dearest* than Mom of the Year—at least at first glance. Take the giant panda . . .

MOM 1 ## Giant Panda Mom: Playing Favorites, Fatally?

One of the good ones: A giant panda mom nuzzles her cub.

She's a "bad" mom because . . . Panda mothers sometimes have two babies—but they rarely raise more than one. "Pandas have good press, but they [can be] bad moms," said Scott Forbes, University of Winnipeg biologist and author of *A Natural History of Families*. Pandas' second offspring, helpless and about the size of a stick of butter, are typically left to their fate in the wild.

Give Mom a break because . . . As the "favored" offspring gets bigger, he or she takes a lot of attention and eats a lot of bamboo. Mothers probably wouldn't be able to provide for two offspring during the eight to nine months until they are fully weaned, Forbes said. "One robust offspring is probably better than two weak ones later on," he said, "so the quality control occurs early on when it's cheap—before you've invested lots of resources."

MOM 2 — Hamster Mom: She Could Just Eat Them Up

Nibble or nurture? A hamster mom approaches one of her offspring.

She's a "bad" mom because . . . Despite their cuddly appearance, hamster moms can be cold-blooded killers—they often eat their own young.

Give Mom a break because . . . She was planning for the best, and now she's just dealing with the rest. Forbes believes hamster moms practice "parental optimism" by creating broods bigger than they may be able to rear. "They don't know how much food is going to be out there," he said. "They create a litter with a few spare offspring to ensure high-quality young even if [food is scarce] or there are developmental defects in others."

MOM 3 | Rabbit Mom: Absentee Parent

Where's Mom? Baby rabbits are often left alone in their burrows.

She's a "bad" mom because . . . Rabbit mothers abandon their young in burrows immediately after birth and return to feed them for only about two minutes daily during their first 25 days. After this brief bout of "drive-by" parenting, young rabbits are left to fend for themselves.

Give Mom a break because . . . Rabbits are tasty, and predators especially enjoy feasting on helpless newborns. Mothers likely avoid their young to keep their underground locations secret—and their precious progeny alive. Though mother-child "quality time" is limited, increased odds of survival may be the greatest gift of all—a little something to remember this Mother's Day.

Brood for breakfast? Long-tailed skinks may eat their own eggs.

MOM 4 | Long-Tailed Skink Mom: Self-Absorbed Pessimist?

She's a "bad" mom because . . . Even though "Mother of the Year" candidates are scarce among reptiles, this lizard is a standout among callous moms. If the skink mother lays a clutch of eggs when there are lots of predators around, she's likely to eat her brood before they hatch.

Give Mom a break because . . . The skink is probably saving her young from an inevitable fate while making herself stronger to ensure another chance to reproduce, Forbes said. "She may decide that because of all the predators, there is no chance of her eggs making it, so she's going to eat them to recycle the nutrients." ■

Longest Polar Bear Swim Recorded

A female polar bear has set the record for the longest consecutive swim—426 miles (687 kilometers) of water, equivalent to the distance between Washington, D.C., and Boston, a new study says.

A female polar bear swam for a record-breaking nine days straight, traveling 426 miles (687 kilometers). The predator made her epic journey in the Beaufort Sea, where sea ice is shrinking due to global warming, forcing mother bears to swim greater and greater distances to reach land—to the peril of their cubs.

> **TRUTH:** PARTIALLY WEBBED FRONT PAWS HELP POLAR BEARS SWIM. THE BEARS MAY USE THEIR BACK PAWS LIKE RUDDERS TO STEER.

Shrinking Ice

The cub of the record-setting bear, for instance, died at some point between starting the swim and when the researchers next observed the mother on land. She also lost 22 percent of her body weight.

"We're pretty sure that these animals didn't have to do these long swims before, because 687-kilometer stretches of open water didn't occur very often in the evolutionary history of the

Polar bears are swimming greater distances because of shrinking sea ice.

polar bear," said study co-author Steven Amstrup, chief scientist for the conservation group Polar Bears International. Amstrup is also the former project leader of polar bear research for the U.S. Geological Survey (USGS), which led the new study.

Another female bear in the study swam for more than 12 days, but appears to have found places to rest during her journey.

Long Swims Deadly for Polar Bear Cubs

Biologists collared 68 female polar bears between 2004 and 2009 to study their movements. Thanks to what study co-author and World Wildlife Fund polar bear biologist Geoff York calls an "accident of technology and design," the researchers noticed data gaps in the bears' whereabouts. The researchers were later able to link the gaps to periods when the bears were at sea.

The scientists examined GPS data for more than 50 female polar bears' long-distance swimming events, defined as swims longer than 30 miles (50 kilometers). This data was then correlated to rates of cub survival. "Bears that engaged in long-distance swimming were more likely to experience cub loss," said study co-author George Durner, a USGS research zoologist in Anchorage, Alaska.

Five of the 11 mothers that had cubs before they began their lengthy swims lost their young by the time the researchers observed them again on land,

A New Food Source

Polar bears usually hunt for seals from blocks of ice when they're out at sea, and they come back to shore when springtime temperatures melt the ice. However, due to climate changes, the ice is melting earlier each year, forcing the polar bears to come ashore sooner, without sufficient food. The bears' early arrival coincides with the time that snow geese are incubating their eggs, and the bears have discovered that these eggs are quite delicious. A recent study on Canada's Hudson Bay by biologist Robert Rockwell has shown that the snow goose population is in no danger of being wiped out, and their eggs may prove to be a valuable and nutritious alternative food source for the bears.

according to the research, presented at the International Bear Association Conference in Ottawa, Canada.

Sea Ice Loss to Continue

Until 1995, summer sea ice usually remained along the continental shelf of the Beaufort Sea, a critical habitat for polar bears due to its rich seal population. Now the sea ice in the Beaufort and Chukchi Seas is retreating from the coast by hundreds of kilometers, Durner said.

In 2010, Arctic sea ice extent was the third lowest on record, part of a long-term trend of ice loss that will continue for decades to come, according to the National Snow and Ice Data Center in Boulder, Colorado.

It's unknown whether the cubs are drowning at sea or whether the metabolically costly act of swimming long distances in nearly freezing water kills them after they reach land.

"So the sort of conditions that contribute to long-distance swimming are likely going to persist in the future, and if cub mortality is directly related to this, then it would have a negative impact on the population," Durner said. ∎

Alligators Surprisingly Loyal to Old Flames

Bringing new meaning to the phrase "see you later, alligator," a new study suggests that female American alligators frequently return to their sexual partners.

In a study of wild alligators in Louisiana's Rockefeller Wildlife Refuge, seven out of ten of the female reptiles studied in multiple years were found to have mated with the same males during the 1995 to 2005

A female American alligator balances a hatchling on her nose.

Alligators-Only Singles Club

The endangered Chinese alligator has quite the set of vocal chords, and researchers believe they have discovered the reason why: The alligators "sing" in order to form singles clubs. In describing the reptilian tunes, a co-author of the study, Xianyan Wang, said, "It sounds like thunder and can travel a long distance." He thought that the songs might be a way for males to attract females, which is usually the case with other animals. Wang and his team found, however, that both male and females reacted similarly to the calls of either gender, which suggests that the singing is a way of finding other alligators so that mating groups can be formed.

mating seasons. Like females of other reptile species, alligators still couple with several males, according to study leader Stacey Lance of the Savannah River Ecology Laboratory in Georgia.

But in an unusual twist, it seems the same female and male find each other during multiple mating seasons. Scientists aren't sure whether the female chooses the male, or the male seeks out the female. Based on the sheer number of alligators in the park, it's incredible "to think that the same two were getting together every year," Lance said. Once the sex is over, however, the "males are out of the picture."

Alligator Baby-Daddies

Over the ten-year period, Lance and colleagues took eggs from a total of 92 alligator clutches—or groups of eggs—and hatched 1,802 babies in the laboratory. "You actually peel the eggshell as they're coming out—they're really quite cute," Lance said.

The team also captured ten female alligators at their nests, drew blood samples, and compared the wild adults' genes with those of the lab-born hatchlings. No male alligators were captured for the study. But knowing the genes of mother and offspring enabled the scientists to piece together the fathers' genes.

The team found that an average of 51 percent of the clutches contained eggs from multiple fathers. But within that number, 87 percent of the clutches had an "obvious" primary male responsible for siring at least half of the babies.

No one knows why female alligators stand by their "men," though it could be for the guarantee of healthy babies, Lance said. "If a female is successful with a certain male, why not stay with him?" she said.

Alligators Do It Like Birds?

The newfound alligator behavior dovetails with the mating habits of birds, the study authors noted. Like alligators, a female bird will pair with a male, then sneak out for liaisons with other males, Lance said.

Referring to birds' apparent dinosaur ancestry, Lance said that "if you think about birds really being modern reptiles in a lot of ways, this suggests that perhaps that behavior we see in birds is more ancient." It also shows that when it comes to sexual behavior, "what you see on the surface is not usually what's going on." ■

> **TRUTH:**
> ALLIGATORS' EGGS HATCH MALE BABIES IN HOT TEMPERATURES AND FEMALE BABIES IN COOLER TEMPERATURES.

How do Giant Pandas Survive?

**A new analysis of panda poop
has finally answered an age-old question:
How do giant pandas survive on a diet
that's 99 percent bamboo
when they have the guts of carnivores?**

Panda Particulars

1. Pandas spend about 12 hours a day eating.
2. Pandas' front paws have enlarged wrist bones that act as thumbs for gripping.
3. The Chinese name for panda is *daxiongmao*, which means "large bear-cat."

Plant-eating animals tend to have long intestines to aid in digesting fibrous material, a trait the black-and-white bears lack. What's more, when the giant panda's genome was sequenced in 2009, scientists found that it lacks the genes for any known enzymes that would help break down the plant fibers found in bamboo and other grasses.

This led researchers to speculate that panda intestines must have cellulose-munching bacteria that play a role in digestion. But previous attempts to find such bacteria in panda guts had failed.

Dropped Clues

The new study looked at gene sequences in the droppings from seven wild and eight captive giant pandas—a much bigger sample than what was used

A giant panda munches on bamboo.

in previous panda-poop studies, said study leader Fuwen Wei, of the Chinese Academy of Science's Institute of Zoology in Beijing.

Wei and colleagues found that pandas' digestive tracts do in fact contain bacteria similar to those in the intestines of herbivores. Thirteen of the bacteria species that the team identified are from a family known to break down cellulose, but seven of those species are unique to pandas.

"We think this may be caused by different diet, the unique inner habitat of the gut, or the unique phylogenetic position of their host," since pandas are on a different branch of the tree of life than most herbivores, Wei said.

The Humans Did It!

Even with help from gut bugs, pandas don't derive much nutrition from bamboo—a panda digests just 17 percent of the 20 to 30 pounds (9 to 14 kilograms) of dry food from bamboo it eats each day. This explains why pandas also evolved a sluggish, energy-conserving lifestyle.

So how and why did pandas became plant-eaters in the first place? Some scientists theorize that, as the ancient human population increased, pandas were pushed into higher altitudes. The animals then adopted a bamboo diet so they wouldn't compete for prey with other meat-eaters, such as Asiatic black bears, in their new homes, said Nicole MacCorkle, a panda keeper at the Smithsonian's National Zoo in Washington, D.C. Pandas will eat meat if it's offered to them, MacCorkle added, but they won't actively hunt for it. ■

"Vampire" Frog
Found in Vietnam

In the treetops of southern Vietnam lives a mysterious tree frog, whose tadpoles are born with black hooked fangs.

The mountain jungles of Vietnam are home to a new breed of "vampire"—a "flying" tree frog dubbed *Rhacophorus vampyrus*. First found in 2008, the 2-inch-long (5-centimeter-long) amphibian is known to live only in southern Vietnamese cloud forests, where it uses webbed fingers and toes to glide from tree to tree.

Adults deposit their eggs in water pools in tree trunks, which protects their offspring from predators lurking in rivers and ponds.

"It has absolutely no reason to ever go down on the ground," said study leader Jodi Rowley, an amphibian biologist at the Australian Museum in Sydney.

TRUTH: THE WORLD'S BIGGEST FROG IS THE SIZE OF A HOUSE CAT.

Fanged Offspring

However, that trick isn't what earned the species its bloodsucking name. Rather, it's the strange curved "fangs" displayed by its tadpoles, which the scientists discovered in 2010.

"When I first saw them by looking through a microscope, I said, 'Oh my God, wow,'" said Rowley, whose research is funded in part by the National Geographic Society's Conservation Trust.

Tadpoles normally have mouthparts similar to a beak. Instead, vampire frog tadpoles have a pair of hard black hooks sticking out from the undersides

The newly discovered "vampire frog" species in Vietnam

of their mouths—the first time such fangs have been seen in a frog tadpole.

What Are Fangs For?

The scientists do not yet know what purpose the fangs serve. However, frogs that raise tadpoles in tree-trunk water holes often feed their young by laying unfertilized eggs as meals. The fangs, Rowley speculated, could help in slicing these open. ■

These Stripes Say Stay Away!

Skunks, badgers, wolverines: A recent study shows these animals' bold color patterns send a powerful message: Danger Ahead!

A skunk's stripes aren't just for style: They may direct predators' eyes straight to the source of the animal's smelly anal spray, which helps tell them to "Stay away!"

A new analysis of data on and pictures of nearly 200 carnivorous mammals—including skunks, badgers, and wolverines—shows that fierce fighters tend to be more boldly colored than more peaceable animals, which tend to use camouflage to stay safe. And those colorations depend on the animals' methods of defense.

TRUTH: SKUNKS HAVE STRIPED SKIN UNDER THEIR FUR.

Real Stinkers, True Biters

Creatures such as skunks, which have long stripes down their body, "tend to be really good at spraying their anal gland secretions—not just dribbling them out," said study leader Ted Stankowich, a biologist at the University of Massachusetts, Amherst. Skunks are known to eject their offensive musk as far as about 10 feet (3 meters).

Other "species that are pretty good at [spraying]—they may not have pure stripes, but their blotches sort of form a stripe down the body."

On the other end, badgers—which bite attackers—often have stripes by their mouths. "We think these stripes may guide predators' attention to the source of danger," said Stankowich. "If you're a badger and your mouth is the source of danger, that's what you want to advertise."

Bold Colors an Alternative to Stinky Spray?

Warning coloration is more typically found in insects, reptiles, and amphibians, such as poison dart frogs. But it is a useful tactic for mammals as well. This nonconfrontational technique for thwarting predators is especially useful for skunks, which prefer not having to spray. Spraying is "costly . . . they're depleting a weapon, using ammunition that might be useful, and it advertises where they are," Stankowich said.

The stripe "strategy" has been a successful one for skunks throughout evolutionary time, Stankowich said, as the same striped pattern has independently evolved multiple times in skunks and related species across the globe.

The zorilla, for example, can spray like a skunk but lives in Africa and is more closely related to weasels than skunks. Yet its fur is striped just like a skunk's, leading the researchers to conclude that the stripes are a good predator deterrent—as is, of course, the ability to spray.

Like skunks, most other mammalian predators use anal gland secretions, but generally in smaller doses to mark territory, Stankowich noted. (Humans and primates lack anal glands.) Skunks and other sprayers, though—finding themselves with a surplus of musk— "may have co-opted it for use as a defense." ∎

Say It, Don't Spray It: Facts About Skunk Sprays

1. A skunk's spray can travel as far as 10 feet (3 meters).
2. Skunks have a limited supply of defensive spray—they can spray only five to eight times before needing to regenerate more.
3. The spray of skunks smells musky, is oily, and is amber colored.
4. The spotted skunk performs a handstand while spraying.
5. Spraying is actually a skunk's last resort—it initially tries to scare off a possible threat by stomping its feet and raising its tail.

Elephant Makes a Stool

First Aha! Moment for Species!

A seven-year-old elephant's sudden burst of insight may redefine elephant intelligence.

In an apparent flash of insight, a young Asian elephant in a zoo turned a plastic cube into a stool—and a tool—a new study says. That eureka moment is the first evidence that pachyderms can run problem-solving scenarios in their heads, then mentally map out an effective solution, and finally, put the plan into action, researchers say.

An Elephant Never Forgets

In 1999, Carol Buckley, founder of the Elephant Sanctuary in Tennessee, reported that resident elephant Jenny and new addition Shirley became excited when they first met. "Shirley started bellowing, and then Jenny did, too. Both trunks were checking out each other's scars. I've never experienced anything that intense without it being aggression," said Buckley. She later discovered that the pair had both performed with the traveling Carson & Barnes Circus—23 years earlier!

Problem-Solving Pachyderm

During the study, seven-year-old Kandula was eager to reach a cluster of fruit attached to a branch that was suspended from a wire, just out of reach. After some apparent thought, the young male rolled a large plastic cube under the branch and stepped up to snatch the treat with his trunk—a feat he repeated several times during multiple days with the cube and with a tractor tire.

The youngest elephant at the National Zoo in Washington, D.C., Kandula had never before been observed moving an object and standing on it to obtain items, and he didn't arrive at his solution by trial and error, said study co-author Diana Reiss, who studies animal intelligence in elephants and dolphins at Hunter College at City University in New York.

Only a few species—such as humans, crows, and chimpanzees—have demonstrated spontaneous insight, the ability to suddenly, mentally figure out the solution to a physical problem, Reiss said.

No Blocked Nasal Passages

Researchers gave Kandula various objects that could have been used to reach the fruit, including sticks that he could have grasped with his trunk to knock the snacks down. That Kandula didn't do this initially puzzled the scientists, until they realized that using sticks in this way would be unnatural for elephants.

Elephants are known to use sticks as tools—as back scratchers, for example—but not when foraging. That's because the mammals rely heavily on the trunks' sense of smell and touch when seeking out food. Holding anything in their trunks would prevent them from effectively feeling and sniffing out dinner, the researchers say.

"It's as if your eyes were in the palm of your hand and I said, 'Pick up this tool and go get that thing.' As soon as you did that, you'd lose your primary sense," explained study co-author Preston Foerder of the City University of New York.

Elephant's "Sudden Revelation"

For several sessions, Kandula just stared at the hanging fruit, ignoring the stick as well as the cube that was nearby. "He did not attempt to use a tool to reach the food for seven 20-minute sessions on seven different days," Hunter College's Reiss said.

> **TRUTH:** ELEPHANTS, ALONG WITH HUMANS, GREAT APES, AND BOTTLENOSE DOLPHINS, ARE THE ONLY ANIMALS THAT CAN RECOGNIZE THEIR OWN REFLECTIONS.

Kandula, an Asian elephant, uses his step stool at the zoo in Washington, D.C.

"And then he finally had what looked to be this sudden revelation, and he headed right over to the block, pushed it in a direct line right underneath the fruit, and stepped right up on it and got the food in one swift movement. "We can't get inside their heads . . . but the fact that he immediately went over to the block suggests that he was imagining [the process] ahead of time," Reiss said.

Primatologist Frans de Waal agreed. "In order to go to another place to go find a tool that is not visibly near the goal, the elephant needs to imagine what he needs, know where to find it, move away from the goal he wants to reach, in order to find the tool, and so on—all of which goes far beyond the usual learning patterns of most animals," said de Waal, of the Yerkes Primate Center at Emory University, in an email.

TRUTH: ELEPHANTS CAN USE THEIR TRUNKS AS SNORKELS.

The finding "is further proof that elephants are right up there with other large-brained animals when it comes to cause-effect understanding and mental problem solving," added de Waal, who wasn't part of the study.

Smarter Than the Average Elephant?

Two older elephants were also tested, but they failed to show the same kind of insightful thinking. "Perhaps they didn't care enough to try to get the fruit, or it could have been age-related," Hunter College's Reiss said.

To be fair, Kandula is an exceptionally inquisitive and intelligent elephant, study co-author Don Moore said. "Among the smart elephants—and all elephants are smart . . . we think Kandula is one of the smarter ones," said Moore, associate director for animal-care sciences at the National Zoo.

Moore said he hopes the discovery will help raise awareness about the plight of Asian elephants, which are endangered. "Studies like this can help people relate more closely to animals because it makes them more like us," he said. "If we can empathize with animals, we are more likely to help conserve them." ■

Biggest Crocodile Ever Caught?

**It took a village. It took three weeks.
And they brought it back, alive.**

In September 2011, the biggest crocodile reportedly ever caught in the Philippines—perhaps the world—was captured alive. The Associated Press reported that the animal was a 21-foot-long (6.4-meter-long) saltwater crocodile.

Villagers threw a fiesta to celebrate the capture of the croc, which a hundred people had to pull by rope from a creek to a clearing, according to

Edwin Cox Elorde, mayor of Bunawan township in the Philippines, stretches his arms over the huge saltwater crocodile.

the Associated Press. The 2,369-pound (1,075-kilogram) animal was suspected of attacking several people and killing two. The animal, named Lolong, survived capture and was held in the village of Consuelo, near Bunawan township.

How Big Is It?

Federal wildlife officials are trying to confirm whether the reptile is the largest crocodile ever captured. *Guinness World Records* lists a 17.97-foot-long (5.48-meter-long), Australian-caught saltwater crocodile as the largest in captivity. Yet herpetologist Brady Barr said, "I'd be surprised if it was truly six meters," adding that a scientist would need to verify the claim.

Alligator biologist Allan Woodward agreed. "There's never been a crocodile longer than approximately 18 feet [5.5 meters]," said Woodward, of the Florida Fish and Wildlife Conservation Commission. "That would be an exceptional jump."

> "Crocodiles learn quickly ... They particularly learn to avoid dangerous situations very quickly. For research purposes, we find that we often have to change capture techniques, because it's very hard to catch them with the same trick twice."
> **James Perran Ross**
> *crocodile researcher,*
> *Florida Museum*
> *of Natural History*

When Crocs Attack

As for whether the crocodile is the perpetrator of the attacks, it's impossible to know unless the animal is killed and cut open, Barr said. Officials did induce the animal to vomit, which produced no human remains. "It's great they didn't kill it," Barr said. "That's commendable [and] very rare."

Barr noted that most crocodile attacks occur because people have depleted croc habitat or prey. In these cases "crocodiles are just turning to the next available food source, and sadly sometimes that happens to be human," Barr said.

Cases of mistaken identity are also possible, when a crocodile thinks a human is a typical prey species. There are also "rogue" animals that purposely kill people, although that's much less common, Barr said. Overall, he said, the "crocodile's not the villain."

A caregiver cleans the giant saltwater crocodile's new home in captivity.

Saltwater crocodiles are considered a species at low risk of extinction. About a thousand of the species roam the Philippines' southern swamplands, where the new catch was found, Philippine wildlife official Glen Rebong told the Associated Press. Though the species isn't under immediate threat, it is protected from hunting by law.

Philippine Secretary of Environment and Natural Resources Ramon Paje told the AP that the crocodile was captured because it was a threat to the community. But he added that the presence of such reptiles is a reminder that the country's remaining habitats need to be protected.

> **TRUTH:**
> **THE AVERAGE LIFESPAN OF A SALTWATER CROCODILE IN THE WILD IS 70 YEARS.**

Bunawan mayor Elorde told the Associated Press that he had plans to make the captured crocodile "the biggest star" in an ecotourism park, which he said would improve people's understanding of the notorious reptiles' role in the environment.

Barr called that "an awesome idea": "These big crocs are a tremendous resource. Australia is a great example—they have a multimillion-dollar tourism industry based around crocodiles. If you do it right, especially for some of these lesser-developed countries . . . it's a great idea." ■

Hibernating Bears Keep Weirdly Warm

Hibernating black bears can dramatically lower their metabolism with only a moderate drop in body temperature, scientists found in 2011. And the science behind it may be helpful to humans.

Black bears have a reputation for being marathon sleepers. The North American mammals generally slumber about five to seven months without eating, drinking, urinating, or defecating, and then they emerge from their dens in the spring none the worse for wear. Scientists have long known that to survive this lengthy fast, the bears drop their metabolism, the chemical process that converts food to energy.

But it was thought that, like most animals, the bears would have to drop their body temperatures to put the brakes on metabolism—each 18-degree Fahrenheit (10-degree Celsius) drop in temperature should equal a 50 percent reduction in chemical activity.

Slow Down

Not so, according to a 2011 study. A black bear in Alaska can lower its

TRUTH:
BLACK BEAR HAIR ISN'T JUST BLACK: THEIR FUR CAN BE BLUE-GRAY, BLUE-BLACK, BROWN, CINNAMON, OR (VERY RARELY) WHITE.

temperature—normally about 91 degrees Fahrenheit (33 degrees Celsius)—by only about 9 to 11 degrees Fahrenheit (5 or 6 degrees Celsius), yet bring its metabolism to an almost grinding halt, at 25 percent the normal rate.

For the study, zoologists at the University of Alaska, Fairbanks, rescued four "nuisance" bears that had been recently captured by the Alaska Department of Fish and Game. Such bears, which live too close to people, are usually euthanized.

Spirit Bears

In the Great Bear Rainforest, a 250-mile-long region down Canada's western coast, live most of the world's spirit bears (also called Kermode bears). Technically, these are North American black bears, but their fur is white. These white black bears account for nearly 1 of every 40 to 100 bears in that region. Scientists believe the unique coloration occurs when both parents pass on a recessive mutation of the MC1R gene (the same one associated with red hair in humans) to their offspring.

The scientists fitted the bears with various devices to record their temperatures, heartbeats, and other factors before placing the animals into artificial dens. The dens were located in an undisturbed forest near Fairbanks that mimicked the animals' natural habitats.

Hibernating Hearts

The scientists also recorded that the bears' heart rates dropped from 55 to 9 beats a minute. During intervals when the bears were not breathing, there were also as many as 20 seconds between beats. That's because when metabolism slows, so does the need for the heart to pump oxygen through the body.

"If we had that kind of longer interval within our heartbeats, we would probably faint," said study co-author Øivind Tøien.

Metabolism Mystery

How the bears' bodies can create the unexpected drop in metabolism is still poorly understood, but Tøien has a few theories. For instance, bears could be like marmots, a hibernating mammal that regulates metabolism by shrinking the mass of its digestive system and then bulking back up when spring comes.

In general, the bear's uncoupling of metabolism from temperature "is yet another amazing thing that black bears can do," noted Bryan Rourke,

Black bears are the most common bear in North America.

a biologist at California State University, Long Beach, who has studied how bears' hearts can withstand hibernation.

Rourke also pointed to the new study's finding that bears can regulate their temperatures to suit individual needs. For instance, a pregnant female black bear in the study did not allow her body temperature to fluctuate as much as other hibernating bears, presumably to protect the fetus.

Helpful to Humans?

Both scientists emphasized that the bear research—published in 2011 in the journal *Science*—could offer practical applications for humans. "A lot of what hibernating mammals can accomplish addresses ways that maybe we could treat things like muscle disease or heart disease," Rourke said.

For instance, understanding how bears can survive with such low amounts of oxygen may help stroke victims who temporarily lose oxygen flow to the brain. Or, unlocking how bears can control their metabolism without dropping their temperatures may provide clues to how people can lose weight.

"Nearly every organ system in the hibernating mammal," Rourke said, "demonstrates some fascinating but contrasting physiology to humans." ■

CHAPTER 4

Creepy
Craw

lies

Ants, worms, spiders, and other slimy, skittery critters aren't the most attractive creatures in the world, but they are some of the weirdest. Scientists are learning that bugs have a whole slew of "superpowers" that are as impressive as they are strange. Wasps have great memories, male spiders give expert massages, snails can survive being eaten alive, and ants can turn into zombies. Read on . . . but you may want to keep that fly swatter handy.

Wasps Can Recognize Faces
Study Says

Are you one of those people who never forgets a face? You've got some company in the animal kingdom—the wasp.

Scientists have discovered that *Polistes fuscatus* paper wasps can recognize and remember one another's faces with sharp accuracy, a new study has found.

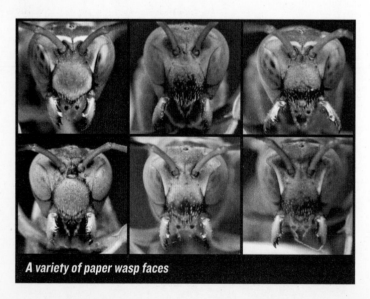

A variety of paper wasp faces

In general, an individual in a species recognizes its kin by many different means. But faces are extremely important to species such as humans, said study co-author Michael Sheehan, a Ph.D. candidate at the University of Michigan in Ann Arbor. "Studies show that when you look at a face, your brain treats it in a totally different way than it does other images," he said. "It's just the way the brain processes the image of a face, and it turns out that these paper wasps do the same thing."

Face-Learning Reaps Rewards

For the study, Sheehan and adviser Elizabeth Tibbetts put wasps of *P. fuscatus* and *P. metricus*—a closely related species with a much less complex social structure—in the long stem of a T-shaped maze. Each wasp in the maze was shown two images of faces of other wasps in the same species—one image to the wasp's left and another to its right. The images "acted like signposts, telling [the subjects] which way to go to get their reward, which in this case was a safety zone," said Sheehan, whose study appeared in the journal *Science*.

Though images and safety-zone locations were constantly changed, "one particular image—face A versus face B—was [always] associated with the safety zone," Sheehan explained. "So they learned, 'If I go to this face, that's good, but the other face does nothing good for me.'"

Repeating the maze experiments using simple shapes or other images instead of faces showed that the wasps learned far more slowly and not as well when faces weren't involved—emphasizing the insects' special response to face recognition.

You Lookin' at Me?

The unique, distinct faces of *P. fuscatus* wasps, as well as the wasps' ability to recognize and remember one another's faces, are likely tied to the insects' multicolony social structure, Sheehan added.

TRUTH: ADULT YELLOW JACKET WASPS BRING FOOD TO THEIR HIVE-BOUND YOUNG, AND IN RETURN, THE YOUNG EMIT SWEET SECRETIONS THAT THE ADULTS CONSUME.

I See Ewe

Sheep, like paper wasps, have also been discovered to possess facial recognition ability. British scientists found that sheep are able to recognize individual faces of at least 50 sheep and can remember them for over two years. "If sheep have such sophisticated facial recognition skills," says Keith Kendrick, author of the study, "they must have much greater social requirements than we thought." Preliminary research also suggests that sheep are able to form mental images of other sheep in their absence. These observations on sheep, paper wasps, and other animals could eventually be useful in developing facial recognition software.

"They have multiple queens and they all want to reproduce—they all want to be the most dominant. So being able to recognize each other helps them understand who's already beaten whom, who has higher ranking in the hierarchy, and this helps to keep the peace. When they aren't able to recognize each other, [as] we've shown before, there was more aggression."

P. metricus wasps, on the other hand, live in single-queen colonies and "don't need to be able to tell each other apart," he said. Not surprisingly, *P. metricus* wasps look alike and do not show the same ability for face-learning, Sheehan said.

Human-Wasp Parallels?

Next, Sheehan hopes to find out how human face perception compares with the ability in wasps. Mammals and wasps have very different eyes, for one thing, and wasp brains are also much smaller and boast far fewer specialized regions.

"We'll be investigating the parallels between primates and wasps," he said. "There are thousands of research papers on face-learning in people, but we're really only beginning to learn about the wasps." ∎

Bumblebee Seeks Warm Flowers
for Heavy Pollination

Scientists have found that bumblebees can be picky pollinators and gravitate toward plants with nice, warm nectar.

A recent study has shown that bumblebees prefer their food warm and learn to locate hotter flowers using color as a cue, scientists say. These findings may have broad implications for the evolution of flowering plants.

Secrets of Attraction

To attract insect pollinators, flowers offer a nutritious reward of nectar and pollen. Now biologists say many flowers may encourage visits by offering a "heat reward" as well. By consuming warmer nectar, bees may save energy they would otherwise have to spend maintaining their own body temperature.

"Bees can raise their body temperature to above 37 degrees Celsius [98.6 degrees Fahrenheit], even if it is just a few degrees above zero outside," said Lars Chittka of Queen Mary College, University of London. "But this is costly, so collecting warm nectar is a clever idea."

> **TRUTH:**
> **BEES CAN BE GREEN, BLUE, OR RED.**

Many flowering plants have features that allow them to increase the temperature of their flowers. The scientists suggest that providing hot meals

To Bee or Not to Bee

According to a new study on bee diversity, bees and the flowers they pollinate are disappearing, which raises concerns about food crops and plant communities that count on animal pollinators to reproduce. Scientists compared a million records on bees from hundreds of sites in the United Kingdom and the Netherlands before and after 1980, and found that bee diversity has decreased at almost 80 percent of the sites. The team also found that plant diversity had suffered a decline, which they believe suggests that the decline in bee and plant diversity is somehow linked.

might be a shrewd evolutionary adaptation for plants, whose own reproduction depends on attracting pollinators.

Heat Seekers

Previous work has shown that some insects are attracted to warmer plants. In tropical rain forests, for example, scarab beetles spend much of their time deep inside flowers capable of generating heat through chemical reactions.

But scientists didn't know if flower temperature is important to more active and widespread pollinators such as bees, which visit each blossom for only a short time. Chittka and his colleagues designed a set of experiments to test the effects of flower temperature on the behavior of pollinating bumblebees.

They laid out a range of flower-shaped feeders holding equal concentrations of sugar solution that varied in temperature. The bees gravitated toward the feeders offering the warmest nectar. The researchers then placed food solutions in pink or purple feeders.

The bees quickly learned to distinguish between colors and concentrated their foraging on the warmer nectar source. The results suggest that the bees learned to use color as a signal of temperature and utilized this information in choosing which flowers to visit.

University of Arizona biologist Daniel Papaj says the study is intriguing. But it remains to be seen whether bees in nature commonly use sensory cues to identify warmer flowers, he notes. If bees do this, Papaj said, "one might expect such behavior to have demographic consequences. For instance, flowers in sunnier microhabitats would be more likely to be pollinated."

Bag of Tricks

The fact that pollinating insects may be choosy about temperature suggests a new explanation for a number of features that help plants keep their flowers warmer than the surrounding environment.

A small number of plant species generate heat through metabolic processes similarly to that of animals. Many others use more passive means to gather and retain warmth. The shape of many flowers—and the ability to track the sun's movements—helps make them efficient collectors of solar energy. Some species even have cone-shaped cells in their petals that focus sunlight, increasing the temperature of the flower.

> **TRUTH:**
> UNLIKE HONEYBEES, BUMBLEBEES CAN STING MORE THAN ONCE BECAUSE THEIR STINGERS ARE SMOOTH AND DO NOT GET CAUGHT IN THE SKIN WHEN THEY FLY AWAY.

"Many of these features had been previously thought to only enhance the color of the flower, or else warm the flower itself up to help its seed develop faster," said study co-author Heather Whitney of the University of Cambridge in England. "Now we know that warming structures could be part of the bag of tricks that flowers have evolved to attract pollinators." ∎

Virus Brainwashes Caterpillars

Just one gene is all it takes for a deadly virus to take over the brains and bodies of caterpillars before it turns them to goo.

Scientists have identified a single gene that allows a caterpillar-brainwashing virus to do its dirty work, a new study says. The virus forces the "zombie" caterpillars to climb trees, where the invader eventually liquefies its hosts' bodies into a dripping goo.

Altered Behavior

"When gypsy moth caterpillars are healthy and happy, they go up into the trees at night to feed on leaves, and then climb back down in the morning to hide [in bark crevices or soil] from predators during the day," said study co-author Kelli Hoover, an entomologist at Penn State University.

But caterpillars infected with a baculovirus—a type of virus that infects invertebrates—are driven to the treetops and reprogrammed to stay there until they meet a doom worthy of a horror film. "When they are infected, as they get sicker they stay up in the trees and die up there," Hoover explained.

> **TRUTH:**
> **A CATERPILLAR HAS MORE MUSCLES THAN A HUMAN.**

The virus "ends up using just about all of the caterpillar to make more virus, and there are other genes in the virus that then make the caterpillar melt. So it becomes a pool of millions of virus particles that end up dropping onto the foliage below where it can infect other moths that eat those leaves."

Master Manipulators

Though such zombie-making viruses were previously known, their genetics have been a mystery. So Hoover and colleagues infected gypsy moth caterpillars with half a dozen different types of baculovirus and placed the bugs in tall bottles with food on the bottom. Viruses that the scientists had determined carried a specific gene, called egt, drove caterpillars to climb to the top of the container and stay there to die.

Researchers then removed egt from some viruses, reinfected the caterpillars, and found that the zombie behavior stopped. When the team inserted the gene into a virus that previously lacked it, the zombie behavior returned.

"Somehow or other, using this gene, the virus is able to manipulate the behavior of the caterpillar to go to the right location in the tree to enhance transmission to new hosts. It's really amazing," Hoover said.

The gene may work by deactivating its hosts' molting hormone, according to the study, published in the journal *Science*. "That would be an advantage to the virus because it keeps the insect in a feeding state, so that they get bigger and bigger and make more and more virus."

Not Berry Funny

In Central and South America, scientists have discovered a parasite that makes ants look like red berries, which tricks birds into eating them. The parasites travel through the birds intact and are then defecated out onto leaves. New ants eat the leaves and are exposed to the parasite, which allows it to keep spreading. As insect ecologist Steve Yanoviak explains it, "No matter how we look at it, somehow that parasite has to infect new colonies, or else it would die . . . So there has to be a mechanism for transport to a new colony."

Natural Enemies

There are many different types of baculovirus, Hoover said, and almost all caterpillar species are infected by one or more of them. But the virus, which is naturally occurring, doesn't greatly impact gypsy moths as a species, Hoover said. Gypsy moth populations are prone to cycles of boom and bust, so when caterpillar numbers are in check, the virus remains so as well.

When gypsy moth invasions grow, the virus may go into outbreak mode, which serves as a natural control mechanism for caterpillar infestations. "This virus probably came to North America when the caterpillars did," Hoover explained. "It's just a natural enemy of the gypsy moth." ■

"Zombie" Ants

Found With New Mind-Control Fungi

A Brazilian rain forest is crawling with mind-controlled zombie ants whose brains are under the sway of a newly discovered fungus that controls their every move.

There's a new mind-controlling fungus in town, but you can rest easy, it only affects ants in a Brazilian rain forest. Originally thought to be a single species, called *Ophiocordyceps unilateralis,* the fungus is actually four distinct species—all of which can "mind control" ants—scientists announced.

Four Fungi

The fungus species can infect an ant, take over its brain, and then kill the insect once it moves to a location ideal for the fungi to grow and spread spores. All four known fungi species live in Brazil's Atlantic rain forest, which is rapidly changing due to climate change and deforestation, said study leader David Hughes, an entomologist at Penn State University.

> **TRUTH:**
> **AN ANT CAN CARRY 50 TIMES ITS BODY WEIGHT.**

Hughes and his colleagues made the discovery after noticing a wide diversity of fungal growths emerging from ant victims, according to the 2011 study in the journal *PLoS ONE.* "It is tempting to speculate that each species of fungus has its own ant species that it is best adapted to attack," Hughes said.

"This potentially means thousands of zombie fungi in tropical forests across the globe await discovery," he said. "We need to ramp up sampling—especially given the perilous state of the environment."

How to Zombify an Ant

The four newly identified "zombie" fungi species use different techniques to spread after infecting an ant, the researchers found. Some of the fungi species create thin "infection pegs" that stick out from a victim's body and infect passing ants, Hughes said.

Other fungus species develop explosive spores on infected ants' bodies. When other ants come near the cadavers, the shooting spores can hit the unwitting passersby, turning them too into zombie ants.

Once they are successfully lodged in a zombie ant's brain, the fungi species "direct" the dying ants to anchor themselves to leaves or other stable places, providing a "nursery" for the fungus. For instance, as the *Ophiocordyceps camponoti-balzani* fungus is about to kill the ant, the insect bites down hard into whatever substance it's standing on. This attachment is so strong that a dead zombie ant can remain stationary even when hanging upside down, the scientists say.

Mmmm. Brains.

Fire ants in South America must watch out for female phorid flies, which have a bizarre reproductive strategy: They hover over fire ants and then inject their eggs into the ants with a needlelike appendage. The egg grows, and the resulting larva generally migrates to the ant's head. The larva lives there for weeks—slurping up the brain and turning the ant into a "zombie," in some cases compelling the ant to march 55 yards (50 meters) away from its colony to avoid attack by other fire ants. Finally, the baby fly decapitates its host and hatches, exiting through the deceased ant's head.

The Cycle Begins Again

The fungus will eventually kill the ant. Once it dies, the fungus rapidly spreads through the body. During the first couple days, though, very little evidence of the fungus is visible from the outside. But a few days after death, white fungus stalks begin to poke through the ant's head. Also noticeable are faint, white, slightly fuzzy fungal growths on the ant's joints.

During later stages of *Ophiocordyceps camponoti-rufipedis* infection, the ant's body takes on a furrier appearance as the fungus rapidly begins to colonize the outside of the ant's body. In the final stages, a long fungus

TURNING INTO A ZOMBIE

1 : *Healthy* Camponotus rufipes *ants* ***scamper across*** *a* ***Brazilian forest floor.***

2 : *To provide a safe place for the growing fungus,* ***infected ants*** *are "directed" to* ***anchor themselves to leaves*** *or other stable places.*

3 : *The* Ophiocordyceps camponoti-rufipedis *fungus* ***rapidly consumes*** *the nutrients inside a zombie ant and begins to colonize the* ***outside of the ant's body,*** *as pictured.*

4 : *The mature fungus stalk, shown* ***growing from a zombie ant's head*** *during the final stage of infection, differs among fungi species.*

4

1

stalk continues to grow from the back of the head, becoming longer and more noticeable. The appearance of each stalk varies with each form of the fungus. For instance, *Ophiocordyceps camponoti-rufipedis* creates just a single stalk while *Ophiocordyceps camponoti-balzani* forms a forked stalk that resembles a reindeer's antlers.

<blockquote>

TRUTH:

A GIANT FUNGUS IN OREGON SPREADS OUT OVER AN AREA THE SIZE OF 20,000 BASKETBALL COURTS.

</blockquote>

Other Insects, Other Zombies

Ants aren't the only zombie-fungi hosts—other insects also fall prey to fungus. Crickets, wasps, and flies can all fall prey to different fungi. For instance, wasps can be infected by a Cordyceps fungus species that hasn't yet been named or formally documented. Fungi of the Cordyceps genus are the products of a tightly evolved arms race between hosts and parasites, study author Hughes noted. That means the fungi are often locked into one type of host—a specialization that might spell doom for fungi species as host species die out.

Much is left to be discovered about these fascinating fungi, and Hughes plans to remedy that—and expects to find many more zombie fungus species in the forests of Brazil. "This is only the tip," he said, "of what will be a very large iceberg." ∎

Cricket Has World's Biggest Testicles

(But Puny Output)

The new title for world's biggest testicles (relative to body weight) goes to the tuberous bushcricket, a type of katydid, according to a new study.

The male tuberous bushcricket, a type of katydid, is now the record holder for the world's biggest testicles relative to his size. But how does their size factor into the bug's fertility?

Production

The sperm-producing organs account for 14 percent of the body mass of males of this bushcricket species. The previous record holder's testicles—belonging to the fruit fly *Drosophila bifurca*—tipped the scales at about 11 percent of its body mass.

"I was amazed by the size of the testes—they seemed to take up the entire abdomen," said study leader Karim Vahed, a behavioral ecologist at the University of Derby in the United Kingdom.

> **TRUTH:**
> CRICKETS CAN SEE DIFFERENT DIRECTIONS AT THE SAME TIME.

But the new heavyweight champion doesn't pack much of a punch. The team was surprised to discover that tuberous bushcrickets have smaller ejaculations than bushcricket species with smaller testicles.

Ideal Study Subjects

For the testicle study, Vahed and colleagues dissected specimens from 21 bushcricket species collected around Europe. The insects are ideal for studying reproductive evolution because of their efficient mating process, Vahed noted. For one thing, the male bushcricket transfers his sperm to the female in a "neat packet" that's easily retrievable by researchers—"whereas in mammals, you'd have to provide some sort of condom to measure the ejaculate," he said.

Likewise the female stores each male's sperm packet in a separate pouch, enabling scientists to count how many times a female has mated in her lifetime. Predictably, the team found that the species whose females mate the most has the males with the biggest testicles, according to the study, published in the journal *Biology Letters*.

But among the 21 bushcricket species, the study showed that, as testicle size increases, ejaculation volume decreases. The discovery runs counter to previous findings in other species—especially mammals. Usually the male with the biggest testicles has more sperm in each ejaculation, thus earning him more tickets in the lottery of fertilizing females, Vahed explained.

Strange Animal Genitalia Facts

1. *Carabid* beetles have only one testicle—and scientists cannot explain why.
2. Rodents with longer penises enjoy a mating advantage.
3. Industrial pollutants are causing polar bears' penises to shrink.
4. Some female ducks have evolved vaginas with clockwise spirals that keep out oppositely spiraled penises of undesirable male ducks.
5. Bats with larger brains have smaller testicles.

Bigger Testicles, More Sperm?

A possible explanation, he said, is that, in societies with promiscuous females, large testicles give males a more plentiful sperm reservoir for multiple matings. Female tuberous bushcrickets mate an average of 11 times in their two-month

life spans. This alternative explanation for large testicles may even make scientists revisit some of their studies on vertebrates, he added. It's usually the other way around.

"It's clear that insects are one of the major types of organisms on planet Earth, [but] the tendency is to draw conclusions from studies of vertebrates and generalize them as if they apply to everything," he said.

Indeed the take-home message for scientists is that mating rate needs to be taken into account when investigating testicle size, according to David Hosken, chair in evolutionary biology at the U.K.'s University of Exeter. Overall the findings are not that surprising, he added via email. "Higher mating rate selects for larger testes, but across other species, sperm competition risk seems to have a greater effect than mating rate."

TRUTH:
FEMALE CRICKETS DO NOT CHIRP; A MALE GENERATES THE CHIRP SOUND WHEN HE RAISES HIS LEFT FOREWING AND RUBS IT AGAINST THE UPPER EDGE OF HIS RIGHT FOREWING.

Bushcricket Titillator Mystery

Next, Vahed plans to shift his focus to "titillators," the hard, penislike part of male bushcricket genitalia that's inserted into the female. These poorly studied—and often spiny—parts may stimulate the female, allow the male to hang on, or both.

And, as it turns out, the tuberous bushcricket isn't quite so well endowed in this arena. The species' parts, he said, "don't seem to be as outlandish as some species that have double sets of spiny titillators." ∎

New "Devil Worm"

Is Deepest-Living Animal

Deep within the earth lives the "devil worm," a species evolved to withstand heat and crushing pressure at extreme depths.

A "devil worm" has been discovered miles under the earth—the deepest-living animal ever found, a new study says. The new nematode species—called *Halicephalobus mephisto* partly for Mephistopheles, the demon of Faustian legend—suggests there's a rich new biosphere beneath our feet.

> "If life arose on Mars and it is still there deep underground, then it may have continued to evolve into something more complex than we are willing to entertain today."
> **Gaetan Borgonie**
> *nematologist, University of Ghent*

Life Down Below

Before the discovery of the signs of the newfound worm at depths of 2.2 miles (3.6 kilometers), nematodes were not known to live beyond dozens of feet (tens of meters) deep. Only microbes were known to occupy those depths—organisms that, it turns out, are the food of the 0.019-inch-long (0.5-millimeter) worm.

"That sounds small, but to me it's like finding a whale in Lake Ontario. These creatures are millions of times bigger than the bacteria they feed on," said study co-author Tullis Onstott, a geomicrobiologist at Princeton University in New Jersey.

Worm Evolved for Harsh Depths

Onstott and nematologist Gaetan Borgonie of Belgium's University of Ghent first discovered *H. mephisto* in the depths of a South African gold mine. But the team wasn't sure if the worms had been tracked in by miners or had come out of the rock.

To find out, Borgonie spent a year boring deep into mines for veins of water, retrieving samples, and filtering them for water-dwelling nematodes. He scoured a total of 8,343 gallons (31,582 liters) until he finally found the worm in several deep-rock samples.

What's more, the team found evidence the worms have been there for thousands of years. Isotope dating of the water housing the worm placed it to between 3,000 and 12,000 years ago—indicating the animals had evolved to survive the crushing pressure and high heat of the depths.

> **TRUTH:**
> THE DEEPEST LIVING FISHES EVER DISCOVERED ON EARTH ARE WHITE SNAILFISH FOUND 4.8 MILES BENEATH THE SURFACE OF THE PACIFIC OCEAN.

"This discovery may not surprise passionate nematologists like Gaetan, but it's certainly shocking to me," Onstott said. "The boundary of multicellular life has been extended significantly into our planet."

Extreme Life on Earth and Elsewhere

Onstott hopes the new devil worm will inspire others to search for complex life in the most extreme places—both on Earth and elsewhere. "People usually think only bacteria could exist below the surface of a planet like Mars. This discovery says, 'Hold up there!' " Onstott said. "We can't negate the thought of looking for little green worms as opposed to little green microbes." ■

Fire Ants Swarm
Form Life Raft

The way that fire ants work together creates a whole new definition of teamwork. When the levee breaks, these creatures stick together.

When a city floods, humans stack sandbags and raise levees. When a fire ant colony floods, the ants link up to form a literal life raft. Now, new research shows exactly how the ants manage this feat.

Ants Ahoy!

Engineering professor David Hu and graduate student Nathan J. Mlot at Georgia Institute of Technology had heard reports of ant rafts in the wild that last for weeks.

"They'll gather up all the eggs in the colony and will make their way up through the underground network of tunnels, and when the flood waters rise above the ground, they'll link up together in these massive rafts," Mlot said.

Together with Georgia Tech systems-engineering professor Craig Tovey, the scientists collected fire ants and dunked clumps of them in water to see what would happen. In less than two minutes the ants had linked "hands" to form a floating structure that kept all the insects safe. Even the ants down below can survive this way, thanks to tiny hairs on the ants' bodies that trap a thin layer of air. "Even when they're on the bottom of the raft, they never technically become submerged," Mlot said.

> **TRUTH:**
> 1,000,000,000,000,000
> (THAT'S ONE QUADRILLION) ANTS
> LIVE ON EARTH.

Form a life raft! Fire ants link up to float on water.

After watching the ants form rafts, the team froze the rafts in liquid nitrogen to study their structures. Ants had linked up using either a "hand to hand" grip, where each ant would grab another's leg, or a mandible-to-leg grip.

Either way, the formation of ant rafts is a delicate business: The maximum strength by which two ants can grasp each other without causing harm is about 400 times their body weight, which is "significantly weaker than ant attachment to other complex surfaces," the paper's authors wrote.

Different Floods, Different Rafts

Ant rafts are quite buoyant, as the researchers found out after attempting, unsuccessfully, to sink one with a twig. During such "perturbations," the ants contract their muscles, which makes the raft temporarily less buoyant but traps air better, preventing drowning.

The scientists created several other flooding scenarios to see how the ants would react. When a cluster of ants is just placed on the water, ants near the top try to leave the pile, Mlot said. But when an escaping ant reaches the edge and

Ant Architects
The fire ant is the only ant species that's been observed forming rafts. But it's been well known that several types of ants can form clusters to create structures similar to towers or bridges.

The ants formed a ball after being swirled in a beaker.

realizes that terra firma is nowhere to be found, "usually the ant will turn around and head back toward the center. By the time she realizes there's an edge, there's already another ant clambering on top of her, forcing that first ant to become part of the bottom layer." In this way, ants that might have tried to survive solo are instead trapped into becoming part of the raft.

Robo-Ant

Research on fire ants could influence robotics. Nathan Mlot, an engineering student at Georgia Tech, explains: "With the ants, we have a group of unintelligent units acting on a few behaviors that allow them to build complex structures... In autonomous robotics, that's what is desired—to have robots follow a few simple rules for an end result."

When the researchers put ants in a dry beaker and gave the container a swirl, the ants rolled into a ball— much like a rolling snowball, Mlot said. It was easy to grasp the ball with a pair of tweezers and submerge the ants, he added. A shimmering layer within the underwater ball shows the edges of an air bubble trapped by the "ant sphere." If the ball had been placed on top of the water, the ants would have instead formed a raft.

Just One Ant

Even a single ant is buoyant in clean water, thanks to rough, waxy hairs that trap air around the insect's body. Soap and other surfactants can reduce the surface tension of water, which in turn reduces an ant's buoyancy.

Despite being denser than water, a single ant can walk on water, thanks to surface tension and hydrophobic (water-repelling) feet. To study how the ant's exoskeleton traps air, the team needed to weigh the ant down. "The only way we could keep an ant underwater was by tying an elastic band around its body and attaching a weight to it," Mlot said.

But surface tension is too weak to support larger objects, which is why it's been a mystery how ant rafts could stay afloat. The answer lies in numbers: When linked together in a raft, the ants' collective water repellency was actually 30 percent higher than that of an individual ant, the researchers found.

Acting Fluid

The Georgia Tech team found that fire ant clusters act like fluids with predictable physical properties: A cluster is a fifth as dense as water but has ten times the surface tension and is ten million times more viscous. Modeling a group of fire ants as a fluid, with each ant representing a molecule, makes describing an ant raft much like describing a drop of oil spreading on the surface of water, the study authors say. Of course, ant "droplets" do not follow totally predictable rules, with individual ants moving randomly within the raft, unlike individual oil molecules.

But when the ant clusters are handled, they don't feel or behave the way water does. In fact, ant clusters feel more like

When the ants are linked together, they behave like a liquid, researchers found.

putty when handled, Mlot said. "You could pick up that ball and it would have the same texture as soft putty. You could give it a squeeze. You could toss it in the air and the ants would stay together." ◼

Snails Survive Being Eaten

by Birds

In the wild kingdom, it's expected that when one animal swallows another one, the digested party dies. But not so with these surprising snails who are able to take a trip through a bird's gullet and come out alive.

Tiny snails can survive being eaten by birds—and the gastropods come out the other end perfectly healthy, a recent study says.

They Will Survive!

Researchers studying feces of wild Japanese white-eye birds had noticed a surprising number of intact snail shells, especially of *Tornatellides boeningi*. This 0.1-inch (0.25-centimeter) snail is common to Hahajima Island about 620 miles (a thousand kilometers) south of Tokyo.

TRUTH: A SNAIL CAN CRAWL ALONG THE EDGE OF A RAZOR WITHOUT CUTTING ITSELF.

So Shinichiro Wada, a graduate student at Tohoku University in Japan, and colleagues fed more than a hundred snails to captive white-eyes and 55 to captive brown-eared bulbuls, another bird known to eat *T. boeningi*. The team found that roughly 15 percent of the snails passed through both bird species' guts alive. One snail even gave birth shortly after emerging—apparently unfazed by its incredible journey.

Branching Out

While the snails are passing through the birds' guts—a process that takes between 30 minutes and two hours—the snails may be inadvertently hitching a ride to new digs. For example, the team found that *T. boeningi* snails in the wild were part of one large genetic group. *T. boeningi* snails whose shells weren't initially found in the white-eye birds' poop, or were found broken, were much more genetically isolated—in other words, they did not tend to move to new locations.

However, there are limits to this mode of travel, the scientists say. Since the birds' digestion is not exactly leisurely, "we are thinking it might be difficult for the snail to migrate over an archipelago," Wada said via email.

Snail Survival a Mystery

The remaining mystery is how the snails manage to survive being eaten. Their small size may prevent their shells from cracking, but the digestive process shouldn't be a comfy ride for any living creature.

Wada said these snails, like many land snails, have the ability to seal their shells' opening with a mucus film called the epiphragm. "This may be a big factor, because their tiny shell aperture and epiphragm would prevent inflow of digestive fluids," he said. ∎

Five Weirdest Bugs

If the natural world were to hold a talent contest, we're confident that these five creatures would be in the running for the weirdest bug trick.

A favorite children's book, *The Very Hungry Caterpillar,* tells the story of how a caterpillar gorges himself before he becomes a butterfly. His huge appetite sets him apart from the rest, and these bugs, worms, slugs, and leeches have their own very special talents as well.

BUG 1 | The Very Toothy Leech

When this leech feeds, large teeth emerge from its anterior sucker.

A new leech king of the jungle, *Tyrannobdella rex*—or "tyrant leech king"—was discovered in the remote Peruvian Amazon in 2010. The up to 3-inch-long (about 7-centimeter-long) leech has large teeth, like its dinosaur namesake *Tyrannosaurus rex*. The *T. rex* leech uses its teeth to saw into the tissues of mammals' orifices, including eyes, urethras, rectums, and noses—the first recognized specimen was plucked from the nose of a girl in Peru's central Chanchamayo Province. What's more, the newfound critter's "naughty bits are rather small," noted study co-author Mark Siddall, curator of invertebrate zoology at the American Museum of Natural History in New York City. "We didn't say the large teeth were compensating for that, but it did come to mind," he quipped.

Can you hear me now? Gryllotalpa vinae *is the loudest of the insects.*

| BUG 2 | **The Very Loud Cricket** |

The mole cricket species *Gryllotalpa vinae* is the loudest of the insects. The critter uses its specialized front legs to dig a megaphone-shaped burrow. Standing inside that dugout, a cricket can chirp loudly enough that humans can hear it nearly 2,000 feet (600 meters) away. Microphones placed 3 feet (a meter) from a cricket's burrow entrance have recorded peak sound levels of 92 decibels, or about the volume of a lawn mower. In fact, using the burrow, *G. vinae* is able to turn an astonishing 30 percent of its energy into sound.

The colorful Borneo ninja slug is green and yellow.

BUG 3 | # The Very "Ninja" Slug

Boasting a tail three times the length of its head, the newly described long-tailed slug is found only in the high mountains of the Malaysian part of Borneo. The new species shoots its mate with "love darts" made of calcium carbonate and spiked with hormones—hence its nickname: ninja slug. Scientists believe this Cupid-like behavior may increase reproduction success.

BUG 4 | # The Very Musical Water Bug

The water boatman's song comes from a very private place.

The 0.07-inch (2-millimeter) water boatman species *Micronecta scholtzi* has an unusual talent: musical genitalia. Engineers and evolutionary biologists in Scotland and France recorded the boatman—which is roughly the size of a grain of rice—"singing" in a tank; the song is loud enough that humans can hear the sounds while standing at the edge of a boatman's pond, but nearly all the sound is lost when the noises cross from water to air. The boatman creates his songs by rubbing his penis against his belly in a process similar to how crickets chirp. Sound-producing genitalia are relatively rare within the animal kingdom, but animals have evolved hundreds of other ways to boost their hoots, howls, and snaps.

Other cockroach species can swim or hiss, but this is the only one that can jump.

BUG 5 | **The Very Bouncy Cockroach**

Prior to the discovery of *Saltoblattella montistabularis* in South Africa's Table Mountain National Park, jumping cockroaches were known from only the late Jurassic period. The newfound species has legs specially built for jumping. "You don't think of cockroaches and cute going in the same sentence, but these guys are really pretty neat," said Quentin Wheeler, the director of Arizona State University's International Institute for Species Exploration. "I like it because it helps clean up the tarnished image of cockroaches"— only a small fraction of the thousands of cockroach species are true pests, he said. "Everyone paints them all with the same nasty brush—this is an example of a cute little animal doing its thing." ∎

Spider "Resurrections"

Take Scientists by Surprise

A group of drowned spiders may have been down, but they weren't out. A few hours later, they baffled scientists by "coming back to life."

Like zombies, spiders in a lab twitched back to life hours after "drowning"—and the scientists were as surprised as anyone. The spiders, it seems, enter comas to survive for hours underwater, according to a study.

The unexpected discovery was made during experiments intended to find out exactly how long spiders can survive underwater—a number of spiders and insects have long been known to be resistant to drowning.

> **TRUTH:**
> MALE WOLF SPIDERS DON'T LIVE FOR MORE THAN A YEAR, BUT THE FEMALES OF SOME SPECIES CAN LIVE FOR SEVERAL YEARS.

Sleeping Spiders

In particular, researchers wanted to determine whether spiders in flood-prone marshes had evolved to survive longer underwater than forest-dwelling spiders can.

Scientists at the University of Rennes in France collected three species of wolf spider—two from salt marshes, one from a forest. The team immersed 120 females of each species in seawater, jostling the spiders with brushes every two hours to see if they responded.

As expected, all the forest wolf spiders (*Pardosa lugubris*) apparently died after 24 hours. The two salt marsh–dwelling species took longer—28 hours for *Pardosa purbeckensis* and 36 hours for *Arctosa fulvolineata*. After the "drownings," the researchers, hoping to weigh the spiders later, left them out to dry. That's when things began to get weird.

Good as New

Hours later, the spiders began twitching and were soon back on their eight feet. "This is the first time we know of arthropods returning to life from comas after submersion," said lead researcher Julien Pétillon, an arachnologist now at Ghent University in Belgium.

Marsh-dwelling *A. fulvolineata,* which took the longest to "die," typically requires about two hours to recover, the researchers discovered. In the wild, the species doesn't avoid water during flooding, while the other salt marsh species generally climbs onto vegetation to avoid advancing water.

The spiders' survival trick depends on a switch to metabolic processes— the processes that provide energy for vital functions in the body—that do not require air, the researchers speculate. Whatever trick these spiders have mastered, Pétillon said, they may not be alone. "There could be many other species that could do this that we do not know of yet." ■

All in the Family

Are males necessary? Maybe not for long, at least in an insect species whose females have begun to develop sperm-producing clones of their fathers inside their bodies.

In the cottony cushion scale—a common agricultural pest that grows to about a fifth of an inch (half a centimeter) long—a new phenomenon has arisen: When some females develop in fertilized eggs, excess sperm grows into tissue within the daughters.

This parasitic tissue, genetically identical to the female's father, lives inside the female and fertilizes her eggs internally—rendering the female a hermaphrodite and making her father both the grandfather and father of her offspring, genetically speaking.

> **TRUTH:**
> AUSTRALIAN LADYBIRD BEETLES ARE USED TO CONTROL COTTONY CUSHION SCALES, WHICH ARE A THREAT TO CITRUS ORCHARDS.

Loving Yourself

Though this new form of reproduction hasn't replaced cottony cushion scale sex, "this parasitic male has taken off like an epidemic in population," said study leader Andy Gardner, an evolutionary theorist at the University of Oxford.

"Once [this trend] gets started, it's going to sweep through the population so all the females carry it. So there's no point for regular males to exist," Gardner added. If the females begin passing on the parasitic male to their offspring, there may eventually be no more need for "baby boy" cushion scales that grow up and produce sperm and fertilize females, Gardner said.

Gardner and the University of Massachusetts's Laura Ross created a population model that predicted how females would respond to this infectious tissue living within their bodies. The results, published in *American Naturalist,* suggest that the females would benefit from the infection, negating the need for males.

Asexuality Still a Mystery

Though the exact timeline of male decline for the species is unknown, Gardner said, in the "long run, I'd expect the [insect species] to suffer because of asexuality."

> "[T]his parasitic male has taken off like an epidemic in population ... Once [this trend] gets started, it's going to sweep through the population so all the females carry it. So there's no point for regular males to exist."
> **Andy Gardner**
> *zoologist, Oxford University*

For instance, though 30 percent of animal species are asexual, in the "vast majority of cases, when we look at species that are asexual, they're relatively recent [evolutionary] events . . . [and they] seem to go extinct quite rapidly. If you mate with yourself, that doesn't generate the sort of adaptive variation that regular sex does."

There are "obvious benefits" of straight-up sex, he said—the offspring get new combinations of genes that can make the species more robust in general.

Overall, it's a mystery why there are so few insect hermaphrodites—only three species are known, all cushion scales. In general, insects are very sexually variable, reproducing in almost every way known to nature—including, in some species, males that can develop from unfertilized eggs.

Confusing matters, cushion scales are "not really hermaphroditic in the usual sense—it's actually two 'individuals' in one body, [which] makes it more intriguing," he said. "We're sort of groping around in the dark just now." ■

Cockroach Brains

May Hold New Antibiotics?

**Cockroaches may make your skin crawl,
but the insects—or, to be exact,
their brains—could one day save your life.**

Put away that roach spray! Cockroaches may come bearing unexpected gifts: The central nervous systems of American cockroaches produce natural antibiotics that can kill off bacteria often deadly to humans, such as methicillin-resistant *Staphylococcus aureus* (MRSA) and toxic strains of *Escherichia coli,* scientists said. Two species of locust tested so far also have the same bacteria-killing molecules in their tiny heads.

Insects Against Infection

The findings suggest that the insect world—which makes up 80 percent of all animals on Earth—may be teeming with new antibiotics, said study co-author Simon Lee of the University of Nottingham in the United Kingdom. Such a discovery is crucial because scientists are scrambling to combat strains of several infectious diseases, including MRSA and *E. coli,* that are resistant to traditional antibiotics.

"It's a promising new lead. We are looking in an unusual place, and to my knowledge no one else is looking there," Lee said. "That's what we need in terms of [finding new] antibiotics, because all the usual places"—such as soil microbes, fungi, and purely synthetic molecules—"have been exhausted."

A Clever Defense

Lee and colleagues dissected the tissues and brains of cockroaches—which "smell as bad as they look," Lee said—and locusts in the lab. The team tested nine separate types of antibacterial molecules found in the insects' brains and discovered that each molecule is specialized to kill a different type of bacteria. This "very clever defense mechanism" allows the bugs to survive in the most dirty of domains, Lee said.

The scientists found the bugs had antibiotics only in their brain tissue, the most essential part of the body, he added. A bug might live with an infected leg, for instance, but a brain infection would almost certainly be fatal.

TRUTH: A COCKROACH CAN LIVE FOR OVER A WEEK WITHOUT A HEAD.

Insect brain–drugs for humans are still years away, Lee said, but there's one hopeful glimmer: When the team added the insect antibiotics to human cells in the lab, there were no toxic effects. ∎

Ladybug Incubators

Enslaved by Wasps

Good child care is tough to find, and one kind of parasitic wasp has solved the problem by brainwashing ladybugs into incubating and then protecting its young.

The parasitic wasp prepares to inject a paralyzed spotted ladybug with an egg.

When the parasitic wasp *Dinocampus coccinellae* is ready to reproduce, it must seek out a nursemaid, a special someone to nurture its young. The bug has come up with a unique recruiting strategy: mind control.

Nurture and Protect

After the wasp finds a spotted ladybug, it paralyzes her with its venom and then injects her with a single egg. In time the egg will hatch into a larva that will develop for a few days and then chew a small hole through the abdomen of the ladybug. The larva will then spin a cocoon between the legs of the ladybug, whose body will rest on top of the cocoon as the larva undergoes metamorphosis.

In a recent study in the journal *Biology Letters,* scientists note that when the ladybugs survive the larva's emergence, the *D. coccinellae* larva then "brainwashes" the bug into defending the vulnerable cocoon from predators, said study co-author Jacques Brodeur, a biologist at the University of Montreal.

"The parasite is taking control of the behavior of its host—that's why we call it bodyguard manipulation," said Brodeur, who worked with Ph.D. student Fanny Maure.

A wormlike wasp larva emerges from its ladybug host.

With his new research, Brodeur said he has solved a mystery that arose during an outdoor stroll in Canada. "It's quite common when you are out in the forest to see ladybugs on top of a cocoon—we were wondering why it was like that," he said.

Not a Death Sentence

Although most parasites eventually kill their hosts, the wasp-infected ladybugs have a more "atypical fate," according to the study—some ladybugs survive their "horrible" ordeal. For example, Brodeur's team observed in field experiments that 30 to 40 percent of the infected ladybugs lived after the young wasp hatched, including some individuals that later laid their own eggs. The host bug can survive because the wasp larva feeds only on tissues that are not crucial for the ladybug's survival, such as fat, the scientists say.

Once a still living ladybug takes up its post on top of the wasp cocoon, the insect will act aggressively toward intruders by flailing its legs, for example. The scientists suspect the twitching behavior comes from venom left in the ladybug's body after the larva emerges and builds its cocoon.

Eaten Alive!

The parasitic jewel wasp injects venom directly into a cockroach's brain to impede its free will. The venom blocks a chemical called octopamine that controls the cockroach's motivation to walk. The wasp then pulls the "zombie" cockroach into its lair and lays an egg in its abdomen. The larva will eventually hatch and eat the living but immobilized cockroach from the inside out. A mature wasp will emerge from the cockroach after about a month.

Two ladybugs take up their post on top of the wasp cocoon.

The length of time that a wasp larva manipulated a ladybug into protecting its cocoon also varied from insect to insect. In some cases the ladybug stayed vigilant until the larva emerged from its cocoon about 20 days later as a young wasp. In other cases the ladybug was under the wasp's sway for just a few days, Brodeur noted.

In the lab, Brodeur and his team placed predatory lacewings into petri dishes that contained either cocoons covered by live ladybugs, other cocoons covered by dead ladybugs, or cocoons that lacked ladybug bodyguards. The results showed that the lacewings were less successful in attacking cocoons being protected by the "brainwashed" ladybugs.

A newborn wasp after emerging from the cocoon.

The team found that wasp larvae that invested more time and energy into controlling their ladybug bodyguards laid fewer of their own eggs as adults than the larvae that did not. It's the first time that scientists have shown a trade-off between host manipulation and fertility, he said. But the parasites are very haughty when choosing their bodyguards: They'll infect only ladybugs. ■

Male Spiders

Massage Their Mates

When a male golden orb-weaver spider wants to get busy and live to tell the tale, he pulls out a special trick: He gives his mate a "back rub," new research shows.

For many spiders, mating is a risky, sometimes deadly, proposition—females of the species are much bigger than the males. The female golden orb-weaver spiders *(Nephila pilipes)* are up to ten times larger than the males. An unlucky suitor might get interrupted in his carnal embrace when a female kicks him off and eats him.

Staying Alive

Male spiders have evolved multiple techniques to avoid this fate, at least before finishing the deed. Male black widows, for instance, pick up scents from females that help the males determine how hungry their love interests are before attempting to mate. Redback spiders in Australia, meanwhile, actually allow themselves to be snacked on to prolong their time with a female.

N. pilipes's strategy involves another trait common among spiders: pedipalps, a pair of appendages that includes male genitals, said study co-author Matjaz Kuntner of the Smithsonian's National Museum of Natural History and the Slovenian Academy of Sciences and Arts.

The male's pedipalps fit perfectly into the female's two genital openings, and he can leave them behind to "plug" the openings. But a male needs to

> **TRUTH:**
> **THE LARGEST SPIDER IN THE WORLD IS WIDER THAN A BASKETBALL.**

While a female golden orb-weaver in the genus Nephila has a snack, a smaller male touches her body.

mate several times in succession to plug both openings and guarantee the female—which can have multiple partners—will have his babies.

To make his mate more receptive between bouts, a male *N. pilipes* will spread silk over her dorsum, or back, in massagelike motions known as mate binding. *N. pilipes* is the only orb-weaving spider known to perform mate binding, Kuntner said, although some spiders from other genera have been known to use the same tactic.

Touch-a, Touch-a, Touch Me?

Researchers had previously theorized that mate binding works because pheromones in the male's silk served to stimulate a female or simply to relax her. But Kuntner's team wasn't so sure.

So Kuntner and colleagues Daiqin Li and Shichang Zhang, from the National University of Singapore, blocked female *N. pilipes*'s sense of touch by covering the spiders' backs with thin layers of superglue. With a second group of spiders, the researchers removed the females' sense of "smell," to test whether they were picking up on chemical cues in the silk.

A Web of Gifts … and Lies?

When male nursery web spiders look for love they present females with a "nuptial gift" of freshly caught prey wrapped in silk. But research has found that only 62 percent of gifts from male nursery web spiders actually contained fresh prey, while the rest contained inedible substitutes. Males without gifts are usually rejected, which may explain why some males who become desperate present a worthless gift such as bits of flower, cotton, or ant husks. However, the females aren't played that easily—a study showed that they disengaged from mating more quickly with the males that gave worthless gifts and moved on.

The team then let the males go at it.

All 17 females that couldn't "smell" calmed down after getting a massage. Females that couldn't "feel" were less likely to let their mates get it on more than once—about 40 percent weren't calmed, and many ate their mates.

Even males that had their spinnerets blocked could avoid being eaten with a well-timed but silkless back rub, the study team observed. The findings show that the silk itself is incidental—the female spiders are probably responding to simply being touched by their suitors. Thus, the paper's authors say, "mate binding . . . could also be more descriptively termed 'mate massaging.' " ■

Alien Wasps Abduct Ants

Drop Them to Get Food

Looking for a way to banish ants from your picnic? According to a new study, wasps have developed a unique method.

In an experiment done with wild insects, scientists in New Zealand recently witnessed the common wasp, an alien invader to the island country, competing for food with the native ant species *Prolasius advenus*. When a wasp approached a mound of food swarming with ants, the wasp would pluck an ant from the pile, fly a ways off, and drop the still-living insect from its jaws.

"To the best of our knowledge, this behavior has never been described before," said study co-author Julien Grangier, a biologist at Victoria University of Wellington.

> ## TRUTH:
> ### MOST WASPS ARE ACTUALLY SOLITARY, NONSTINGING VARIETIES.

Wasps Bigger, but Not Badder

Common wasps *(Vespula vulgaris)* are native to North America but were accidentally introduced to New Zealand in the 1970s. The wasps eat other insects and nectar, capturing live prey or scavenging. Grangier and colleague Philip Lester had suspected that the alien wasps were competing with native ants for scarce protein sources in New Zealand beech forests.

This suspicion led the pair to establish an experiment in which ants and wasps were presented with samples of high-protein food: little chunks of tuna fish. The samples were placed at 48 stations in a natural beech tree forest, with cameras set up near each one. Both wasps and ants visited 45 of the 48 stations, and the cameras recorded 1,295 interactions between the insects.

In the vast majority of instances, the wasps and ants avoided or ignored each other. However, the researchers documented 341 cases when the ants were aggressive toward the wasps, charging at the larger bugs, biting them, or spraying them with formic acid, a natural defense mechanism.

In just 90 encounters the wasps were the aggressors, including 62 cases of ant dropping. The researchers suspect the other 28 times were ant-dropping attempts that the wasps fumbled. "It was a surprise to see that ants, being 200 times smaller than wasps, can be serious competitors with them," Grangier said.

Ant Acid Behind Wasp Behavior?

Most of the time, the wasps' ant-dropping behavior was unprovoked, with ants being simply grabbed and flown away. In a few instances the ants were unruly before they were grappled and carried off.

Picnic Pests

Western yellowjacket wasps, which were accidentally introduced to Hawaii during the 1900s from the U.S. Pacific Northwest, have an astonishingly diverse diet. Adult yellowjackets consume only nectar, but they kill or scavenge prey to provide necessary protein to their growing offspring. "They basically just carry it in their mandibles—you see them flying with their balls of meat," said lead study author Erin Wilson. Wilson and her team found that their prey spans 14 taxonomic groups of animals, including tree lice, spiders, rats, and geckos.

The team argues that the acid defense may be why the wasps "ant drop" rather than just killing the smaller insects outright. "By not crushing ants and dropping them away as fast as possible, wasps just protect themselves, avoiding further contact with this harmful substance," Grangier said. ∎

CHAPTER 5

Explai
Unexp

ning the
lained

You are about to enter a chapter filled with legendary monsters, mysterious disappearances, and conspiracy theories. These unexplained mysteries have become obsessions, possessing the minds of those who cannot abandon their quest for the answers. What happened to Amelia Earhart over the Pacific? Are there secrets hidden at Area 51? Does Bigfoot roam the forests of North America? Is the truth really out there? Come with us and find out.

Titanic Discovered

During Secret Cold War Navy Mission

A "Top Secret" hunt for two sunken submarines gave this explorer the chance of a lifetime: to find the final resting place of the *Titanic*.

The 1985 discovery of the *Titanic* stemmed from a secret United States Navy investigation of two wrecked nuclear submarines, according to the oceanographer who found the infamous ocean liner. Pieces of this Cold War tale have been known since the mid-1990s, but more complete details are now coming to light, said *Titanic*'s discoverer and National Geographic explorer-in-residence, Robert Ballard.

"The Navy is finally discussing it," said Ballard, an oceanographer at the University of Rhode Island in Narragansett and the Mystic Aquarium and

The bow railing of the submerged Titanic

Institute for Exploration in Connecticut. Ballard met with the Navy in 1982 to request funding to develop the robotic submersible technology he needed to find the *Titanic*.

Secret Mission

Ronald Thunman, then the deputy chief of naval operations for submarine warfare, told Ballard the military was interested in the technology—but for the purpose of investigating the wreckage of the U.S.S. *Thresher* and U.S.S. *Scorpion*.

Since Ballard's technology would be able to reach the sunken subs and take pictures, the oceanographer agreed to help out.

He then asked the Navy if he could search for the *Titanic,* which was located between the two wrecks. "I was a little short with him," said Thunman, who retired as a vice admiral and now lives in Springfield, Illinois. He emphasized that the mission was to study the sunken warships.

Titanic Debris

The wreck of the *Titanic* lies in two sections, about 2,000 feet (609 meters) away from each other. In between them is a huge debris field, strewn with pots, pans, plates, and bottles of champagne, as well as personal items: hairbrushes, hand mirrors, a ceramic doll's head, and a pair of boots.

Once Ballard had completed his mission—if time was left—Thunman said, Ballard could do what he wanted, but never gave him explicit permission to search for the *Titanic*.

Ballard said Navy Secretary John Lehman knew of the plan. "But the Navy never expected me to find the *Titanic,* and so when that happened, they got really nervous because of the publicity," Ballard said.

Sunken Subs

The *Thresher* and *Scorpion* had sunk in the North Atlantic Ocean at depths of between 10,000 and 15,000 feet (3,000 and 4,600 meters). The military wanted to know the fate of the nuclear reactors that powered the ships, Ballard said.

This knowledge was to help determine the environmental safety of disposing of additional nuclear materials in the oceans. The Navy also wanted to find out if there was any evidence to support the theory that the *Scorpion* had been shot down by the Soviets.

Ballard's data showed that the nuclear reactors were safe on the ocean bottom and were having no impact on the environment, according to Thunman.

The data also confirmed that *Thresher* likely had sunk after a piping failure led to a nuclear power collapse, he added. Details surrounding the *Scorpion* are less certain.

A catastrophic mishap of some sort led to a flooding of the forward end of the submarine, Thunman said. The rear end remained sealed and imploded once the sub sank beneath a certain depth. "We saw no indication of some sort of external weapon that caused the ship to go down," Thunman said—dismissing the theory that the Russians torpedoed the submarine in retaliation for spying.

Titanic Facts:

1. *Titanic* was massive—883 feet long, 92 feet wide, and 175 feet tall.
2. Touted as "unsinkable," the ship first set sail on April 10, 1912, with 2,223 people on board.
3. The *Titanic* carried lifeboats for only 1,178 people.
4. At 11:40 p.m. on April 14, the first mate spotted a dangerous iceberg. The ship's crew attempted to avoid it, but the *Titanic* was moving too fast and struck the iceberg.
5. The ship sank at 2:20 a.m. on April 15, 1912.
6. The wreck of *Titanic* was found some 13 miles east of its last reported position.

Trail to Titanic

While searching for the sunken submarines, Ballard learned an invaluable lesson on the effects of ocean currents on sinking debris: The heaviest stuff sinks quickly. The result is a debris trail laid out according to the physics of the currents.

With just 12 days left over in his mission, Ballard began searching for the *Titanic*, using this information to track down the ocean liner. He speculated that the ship had broken in half and left a debris trail as it sank. "That's what saved our butts," Ballard said. "It turned out to be true."

The explorer has since used a similar technique to find other sunken ships and treasures, including during his expeditions to the Black Sea. Were these expeditions also part of top-secret missions? After all, the Black Sea is in the volatile Middle East.

"The Cold War is over," Ballard said. "I'm no longer in the Navy." ∎

Bigfoot Discovery
Declared a Hoax

Two men claimed that they had a Bigfoot encounter in the woods of Northern Georgia, but was the whole thing an elaborate hoax?

Alas, the search for Bigfoot continues. No evidence has emerged to support claims made by two men who said they found the corpse of a seven-foot-tall (two-meter-tall) Bigfoot—an apelike creature of North American legend—in the woods of northern Georgia.

Critics declared the men's story a bold hoax after the pair refused to show the body and following the disclosure that genetic tests from the alleged remains revealed only human and opossum DNA.

Stumbling Upon Sasquatch

Matt Whitton and Rick Dyer spoke to a packed room of reporters in Palo Alto, California, about their discovery. Joining them on stage was controversial Bigfoot hunter Tom Biscardi. Whitton told a compelling story of how he and Dyer found the body of the dead Sasquatch—as the creature is also called—next to a stream while hiking in the Georgian woods in late spring.

Whitton said he stood guard by the body for nine hours while Dyer went back to get a truck. When Dyer returned, the pair dragged the hairy 500-pound (230-kilogram) corpse through the woods to the truck—all while being shadowed by three live Sasquatch.

"As we were bringing it out, they were paralleling us," said Whitton, a Georgia police officer on administrative leave. Whitton said that after reaching their truck, they refrigerated the Bigfoot body and soon after contacted Biscardi.

Weak Evidence

At the press conference, the self-proclaimed "best Bigfoot hunters in the world" declined repeated requests to display the Sasquatch remains. Instead, they handed out photographs purportedly showing the creature's mouth and tongue, and a blurry image of a hairy figure strolling through the woods.

Reporters and other Bigfoot investigators were underwhelmed by the group's evidence. "When I first heard about this, I was optimistic and hopeful," said Jeff Meldrum, an anthropologist and Bigfoot investigator at Idaho State University. "But when I heard [Tom Biscardi] was involved, that optimism quickly evaporated."

Within the community of amateur and professional Bigfoot hunters, Biscardi has a "reputation of ill repute," Meldrum said. Meldrum is also extremely skeptical about the authenticity of a photograph released in the days leading up to the press conference showing what appears to be a hairy corpse in a refrigerator.

"It looks like a heap of costume fur. It doesn't look like natural hair," Meldrum said. "The gut pile looks like it was dumped on there just for effect."

> "They believe in a race of giants, which inhabit a certain mountain off to the west of us ... They are men stealers. They come to the people's lodges at night when the people are asleep and take them ... to their place of abode without even waking ... If the people are awake, they always know when they are coming very near by their strong smell that is most intolerable."
>
> **Elkanah Walker**
> American missionary to the Spokane Indians in Washington State, 1840

DNA Results

Casting further doubt on the group's claim are mixed DNA results from the purported body. The DNA sample was analyzed by Curt Nelson, a molecular biologist at the University of Minnesota, who described it as a mixture of human and opossum.

Biscardi's "suggestion was that the tissue sample was from the intestine of the animal, and that the animal had eaten an opossum," Nelson told National Geographic News. "That seems improbable to me."

Jason Linville is a forensics expert at the University of Alabama at Birmingham who was not involved in the DNA analysis. If the group had instead sent in hair samples from the body, it would have been relatively simple to confirm that it belonged to an unidentified primate species, Linville said.

"In theory, you could analyze that DNA and it would come up as something that didn't quite match human and didn't quite match primate, but was something pretty close to it," Linville said.

Big Controversy

Matthew Moneymaker is the president of the Bigfoot Field Researchers Organization, an international network of Bigfoot investigators. Moneymaker called the press conference an elaborate "profiteering scam" engineered by Biscardi.

"They know there's tremendous interest in seeing photographs of [Bigfoot], and they're trying to get people to pay to see hoaxed photos," he said. Moneymaker's organization tracks Bigfoot news in the media, and he says Biscardi really scored with his latest exploit.

"There's been at least a thousand stories in newspapers across the world," Moneymaker said. "Before this, the highest record was about 200 articles in newspapers." Moneymaker predicts that the mass exposure could actually hurt Biscardi in the long run.

Modern Sightings

The Bigfoot Field Researchers Organization maintains a national database of Bigfoot sightings that dates back to the 1980s with sightings reported as recently as 2011. Reports have come in from all over the world—from Malaysia and the Himalaya with the bulk of reports coming from Canada and the United States.

"Now he's really a famous con man," Moneymaker said. "He was a con man known in Bigfoot circles for years, and now it won't be long before everybody knows it."

Biscardi, Whitton, and Dyer remain undeterred, however. The trio said they plan to conduct an autopsy of their Bigfoot corpse in the near future. "I want to get to the bottom of it," Biscardi said. "What I seen, what I touched, what I felt, and what I prodded was not a mask that was sewn on a bear hide, OK?" ■

Chupacabra Science

How Evolution Made a Monster

Tales of a mysterious monster abound in Mexico, the U.S., and even China since the mid-1990s, when the chupacabra was first "sighted" in Puerto Rico. Now, scientists say that evolutionary theories can explain the truth.

Flesh-and-blood chupacabras have allegedly been found as recently as 2010—making the monsters eminently more accessible for study than, say, the Loch Ness Monster or Bigfoot. The legend had its origins in Puerto Rico, but sightings have spread to southwestern United States and Mexico. In many recent cases in Mexico, the monsters' corpses have turned out to be coyotes suffering from very severe cases of mange, a painful, potentially fatal skin disease that can cause the animals' hair to fall out and skin to shrivel, among other symptoms.

For some scientists, this explanation for supposed chupacabras is sufficient. "I don't think we need to look any further or to think that there's yet some other explanation for these observations," said Barry O'Connor, a University of Michigan entomologist who has studied *Sarcoptes scabiei,* the parasite that causes mange.

Animal Attacks

In winter 1995, Puerto Rico was besieged with reports of alleged chupacabra attacks:

1. In Orocovis, eight sheep, completely drained of blood, were found with puncture wounds.
2. In Guanica, chickens and cows died of blood loss with single puncture wounds to the neck.
3. In Torrecilla, Baja, a woman found a chicken dead of perforations in the neck, her cat dead with its genitals missing, and the throats of her guinea pigs slit.

A Texas woman holds a photo of what she believes is a chupacabra.

Mangy Varmints

Likewise, wildlife-disease specialist Kevin Keel has seen images of an alleged chupacabra corpse and clearly recognized it as a coyote, but said he could imagine how others might not. "It still looks like a coyote, just a really sorry excuse for a coyote," said Keel, of the Southeastern Cooperative Wildlife Disease Study at the University of Georgia.

"I wouldn't think it's a chupacabra if I saw it in the woods, but then I've been looking at coyotes and foxes with mange for a while. A layperson, however, might be confused as to its identity."

The Problem With Parasites

Sarcoptes scabiei also causes the itchy rash known as scabies in humans. In humans and nonhuman animals alike, the mite burrows under the skin of its host and secretes eggs and waste material, which trigger an inflammatory response from the immune system.

In humans, scabies—the allergic reaction to the mites' waste—is usually just a minor annoyance. But mange can be life threatening for canines such as coyotes, which haven't evolved especially effective reactions to *Sarcoptes* infection.

The University of Michigan's O'Connor speculates that the mite passed from humans to domestic dogs, and then on to coyotes, foxes, and wolves in the wild.

His research suggests that the reason for the dramatically different responses is that humans and other primates have lived with the *Sarcoptes* mite for much of their evolutionary history, while other animals have not.

"Primates are the original hosts" of the mite, O'Connor said. "Our evolutionary history with the mites helps us to keep [scabies] in check so that it doesn't get out of hand like it does when it gets into [other] animals." In other words, humans have evolved to the point where our immune systems can neutralize the infection before the infection neutralizes us.

The mites too have been evolving, suggested the University of Georgia's Keel. The parasite has had time to optimize its attack on humans so as not to kill us, which would eliminate our usefulness to the mites, he said.

The Trouble With Mange

In nonhuman animals, *Sarcoptes* hasn't figured out that balance yet. In coyotes, for example, the reaction can be so severe that it causes hair to fall out and blood vessels to constrict, adding to a general fatigue and even exhaustion.

Since chupacabras are likely mangy coyotes, this explains why the creatures are often reported attacking livestock. "Animals with mange are often quite debilitated," O'Connor said. "And if they're having a hard time catching their normal prey, they might choose livestock, because it's easier." As for the blood-sucking part of the chupacabra legend, that may just be make believe or exaggeration. "I think that's pure myth," O'Connor said.

Monster "Facts"

Name: Chupacabra
Aliases: El Vampiro de Moca, Goatsucker
Description: Varies. At times said to resemble a large dog or a spikey-backed lizard with big, red eyes. Stands anywhere from 4 to 5 feet (1.2 to 1.5 meters) tall.
Methods: Kills animals by sucking their blood.
Victims: Livestock, including goats, chickens, cows, and horses.
Sightings: Puerto Rico; Mexico; southwestern United States; Miami, Florida.

"Evolution" of a Legend

Loren Coleman, director of the International Cryptozoology Museum in Portland, Maine, agreed that many chupacabra sightings—especially the more recent ones—could be explained away as appearances by mangy coyotes, dogs, and coyote-dog hybrids, or coydogs.

"It's certainly a good explanation," Coleman said, "but it doesn't mean it explains the whole legend." For example, the more than 200 original chupacabra reports from Puerto Rico in 1995 described a decidedly uncanine creature.

"In 1995, chupacabras was understood to be a bipedal creature that was three feet [about a meter] tall and covered in short gray hair, with spikes out of its back," Coleman said.

But, as if in a game of telephone, the description of the chupacabra began to change in the late 1990s due to mistakes and mistranslations in news reports, he said. By 2000 the original chupacabra had been largely replaced by the new, canine one. What was seen as a bipedal creature now stalks livestock on all fours.

"It was actually a big mistake," Coleman said.

"Because of the whole confusion—with most of the media reporting chupacabras now as dogs or coyotes with mange—you really don't even hear any good reports from Puerto Rico or Brazil anymore like you did in the early days. Those reports have disappeared and the reports of canids with mange have increased."

Monkeys or Movies?

So what explains the original chupacabra myth?

One possibility, Coleman said, is that people imagined things after watching or hearing about an alien-horror film that opened in Puerto Rico in the summer of 1995. "If look at the date when the movie *Species* opened in Pue Rico, you will see that it overlaps with the first explosi of reports there," he said. "Then compare the images Natasha Henstridge's creature character, Sil, and you w see the unmistakable spikes out the back that match those of the first images of the chupacabras in 1995."

Another theory is that the Puerto Rican creatures were an escaped troop of rhesus monkeys on the island, which often stand up on their hind legs. "There was a population of rhesus monkeys being used in blood experiments in Puerto Rico at the time, and that troop could have got loose," Coleman said.

"It could be something that simple, or it could be something much more interesting, because we know that new animals are being discovered all the time." ∎

Exclusive
Area 51 Pictures
Secret Plane Crash Revealed

No word yet on alien starships, but now that many Cold War–era Area 51 documents have been declassified, veterans of the secret U.S. base are revealing some of their secrets.

After a rash of declassifications, details of Cold War workings at the Nevada base, which to this day does not officially exist, are coming to light—including never before released images of an A-12 crash and its cover-up.

Area 51 was created so that U.S. Cold Warriors with the highest security clearances could pursue cutting-edge aeronautical projects away from prying

In an undated picture, a mock-up of the A-12 spy plane sits perched upside down on a testing pylon at Area 51.

eyes. During the 1950s and '60s, Area 51's top-secret OXCART program developed the A-12 as the successor to the U-2 spy plane.

Secret Weapon

Extensive testing was performed to reveal how visible, or invisible, the A-12's design was to radar. Area 51 staff had to regularly interrupt such tests and hurry prototypes into "hoot-and-scoot sheds"—lest they be detected by Soviet spy satellites.

The A-12 was about 93 percent titanium, a material then unheard of for aircraft design. Most of the men who built the craft are still wondering today where that metal came from—some sources say it was secretly sourced from inside the U.S.S.R.

Nearly undetectable to radar, the A-12 could fly at 2,200 miles an hour (3,540 kilometers an hour)—fast enough to cross the continental United States in 70 minutes. From 90,000 feet (27,400 meters), the plane's cameras could capture foot-long (0.3-meter-long) objects on the ground below.

But pushing the limits came with risks—and a catastrophic 1963 crash of an A-12 based out of Area 51. A rapid government cover-up removed nearly all public traces of the wrecked A-12—pictured publicly for the first time on National Geographic Daily News, thanks to the CIA's recent declassification of the images.

> **TRUTH:** YOU CAN SEE SATELLITE PICTURES OF AREA 51 ON THE INTERNET BY ENTERING ITS GPS COORDINATES (37°14′36.52″N, 115°48′41.16″W) INTO GOOGLE EARTH.

The Pilot Speaks

Things went horribly wrong for test pilot Ken Collins (flying under his Area 51 code name Ken Colmar) when testing the plane's subsonic engines at low altitude. At 25,000 feet (7,620 meters), "the airplane pitched up and went up and got inverted and went into a flat incipient spin," Collins has said.

From such a position, "you just can't recover. So I thought I'd better eject, so I ejected down, because I was upside down." U.S. officials later asked Collins to undergo hypnosis and treatments of sodium pentothal (a "truth drug") to be sure he relayed every detail of the incident truthfully and correctly.

DECLASSIFIED!

1: *Suspended upside down, a **titanium A-12 spy-plane** proto-type is prepped for radar testing at Area 51 in the late 1950s.*

2: *An aerial photo shows a **massive rapid-response team** at the site of the secret crash.*

3: *A crane hoists A-12 debris (right) onto a flatbed truck. Part of an **engine nacelle and an exhaust ejector** are visible at left.*

4: *Before the cleanup, after pilots from Area 51 had reported that the wreck in Utah was still identifiable, **crews quickly covered all large pieces** with tarps.*

1

Crash Cover-Up

After pilot Ken Collins had parachuted to the ground, he was stunned to be greeted by three civilians in a pickup, who offered to give him a ride to the wreckage of his plane. Instead, Collins got them to give him a ride in the opposite direction, by telling them the plane had a nuclear weapon on board—a prearranged cover story to keep the Area 51 craft a secret.

Soon a team of government agents appeared to direct a complete cleanup—and cover-up—operation. "There was some debate over whether to dynamite the large sections of wreckage, to make identification by unauthorized personnel more difficult," said independent aerospace historian Peter Merlin.

By the next morning, recovery crews had begun loading the wreckage on trucks for the return trip to Area 51 in Nevada. No one else approached the wreck site or even learned of the crash during the next half century.

Secret No Longer!

Area 51 remained hidden for decades until 1988, when a Soviet satellite photographed the base. The photographs were published, and the secret was out!

National Security

"At the time of the crash, the OXCART program was a very closely kept secret, and any exposure of it—such as through a crash that got publicized—could have jeopardized its existence," CIA historian David Robarge said.

"If U.S. adversaries used that disclosure to figure out what the program was about, they might have been able to develop countermeasures that would make the aircraft vulnerable. The U.S. government had to make sure that no traces of the 1963 crash might be found and give hostile powers insights into the engineering and aeronautical advances the program was making."

Decision to Declassify

Today that secrecy has outlived its use, according to the CIA's Robarge, explaining why the crash photos have been declassified. "CIA records managers review [information requests] case-by-case to determine whether the information sought is still sensitive on national-security grounds. In their judgment, the photos of the 1963 crash no longer are, and so they were declassified and released," Robarge told National Geographic News.

"In 2007 the CIA declassified over a thousand documents related to the OXCART program and published an unclassified history of it in conjunction with the acquisition from the Air Force of one of the nine remaining A-12 airframes," now on display at CIA headquarters, Robarge added.

What Remains

Aerospace historian Peter Merlin, who has examined this crash site and several others involving secret aircraft, said he's pieced together at least part of the cover-up story. "The A-12's fuselage and wings were cut apart with blowtorches and loaded onto trucks along with the tails and other large pieces," he said. "Smaller debris was packed in boxes."

Merlin's research into recently declassified documents on the OXCART project unearthed a memorandum that reported that all traces of the plane had been removed from the crash scene in 1963. "My experience with crash sites, however, is that there is always something left."

And in fact recent investigations of the site have turned up parts of the plane's wing structure as well as cockpit remnants still bearing the stamp "skunk works"—the covert department of the defense contractor Lockheed, which worked on the plane.

Unanswered Questions

Though the CIA has released some photos of the incident, officials remain mum about exactly who was involved in the cover-up and how it was carried out. "There's nothing I can tell you about how [this or] any other incidents were or are handled," CIA historian David Robarge said.

Today, experts at Area 51 are likely working on the next generation of aircraft. But don't expect any information to emerge for several decades—despite the recent declassifications, CIA's Robarge still won't confirm the base exists. "Sorry," he said, "I can't say anything about it." ■

Name That Plane!

Area 51 has developed several of the most successful spy craft with some of the most interesting names:

1. **The Suntan:** A plane designed to be a successor to the U-2, which could fly at speeds up to 2,000 mph 3,219 km/h). It was fueled by liquid hydrogen.
2. **The Bird of Prey:** Named for a class of ships from *Star Trek*. A bomber with stealth technology but unstable at low speeds.
3. **Tacit Blue:** One of the first successful attempts at creating stealth aircraft. Had an odd, whalelike shape, inspiring the nickname "Shamu." Designed to fly low over battle operations as a reconnaissance vehicle.

Kraken Sea Monster Account

"Bizarre and Miraculous"

Fossils in a prehistoric ocean graveyard have raised the theory of a kraken who moonlights as an artist. But are they more easily explained?

> **TRUTH:**
> MODERN-DAY GIANT SQUIDS CAN GROW AS LONG AS 43 FEET IN LENGTH.

Are fossils revealing the existence of a giant prehistoric squid's art? The fossils in question are about 350 miles (560 kilometers) northwest of Las Vegas, in Nevada's Berlin-Ichthyosaur State Park—a seafloor at the time the bones were deposited, some 200 million years ago.

The fossils are circular vertebral discs, or backbones, that once belonged to *Shonisaurus populatis*, a species of ichthyosaurs. Based on the bones' sizes, scientists estimate the ichthyosaurs grew to lengths of 49 feet (15 meters) or longer.

Portrait of an Artist?

During a recent family trip to the fossil site, Mark McMenamin, a paleontologist at Mount Holyoke College in Massachusetts, noticed that some of the vertebrate fossils appeared to be neatly lined up into double rows.

Struck by the orderly arrangement of the bones, McMenamin came up with a remarkable idea for how they came to be that way, which he presented at a meeting of the Geological Society of America. McMenamin's hypothesis:

A giant squid or octopus hunted and preyed on the ichthyosaurs and then arranged their bones in double-line patterns to purposely resemble the pattern of sucker discs on the predator's tentacles.

According to a press release detailing McMenamin's hypothesis—titled "Giant Kraken Lair Discovered"— "the vertebral disc 'pavement' seen at the state park may represent the earliest known self portrait . . . I think that these things were captured by the kraken and taken to the midden and the cephalopod would take them apart," McMenamin said in the statement. The kraken, he said, "was either drowning [the ichthyosaurs] or breaking their necks."

Things Fall Apart

Paul "P.Z." Myers is an evolutionary biologist at the University of Minnesota, Morris and the author of Pharyngula, a science blog that is partially funded by the National Geographic Society. Myers called McMenamin's hypothesis a "bizarre and miraculous story" and said his evidence is "weirdly circumstantial." The fossil arrangement "is not surprising," Myers said. "It doesn't take an artist octopus to do it." One could imagine, Myers said, that as ichthyosaurs died and their bodies rotted, their vertebral discs fell apart. The bones "are taller than they are wide, so they're just going to flop over to one side or the other and can just happen to fall into two parallel rows," which then get preserved as fossils, he said.

"Fun to Think About"

McMenamin's kraken ideas have received media attention partly because they were presented at a scientific conference, but that's no sign that a hypothesis is widely accepted or considered scientifically plausible, Myers added. Scientific meetings "are where scientists go to talk with their peers and discuss preliminary data, so they naturally have fairly lax standards," Myers said.

Ryosuke Motani, a paleontologist at the University of California, Davis, who has also conducted research at Berlin-Ichthyosaur State Park, was equally skeptical of McMenamin's idea. "It's fun to think about," Motani said, "but I think it's very implausible." Motani proposed an alternative hypothesis for how the bones came to be arranged the way they are. "These bones are disc-shaped, so when they're disarticulated after rotting, they lay flat on the seafloor and can get gathered up and packed together by ocean currents," he said.

"This particular specimen [that McMenamin focused on] has two rows. But I've seen others that have three rows . . . It's natural that the bones get arranged like that." ∎

King Tut Mysteries Solved:

Disabled, Malarial, and Inbred

King Tut may be seen as the golden boy of ancient Egypt today, but a DNA analysis reveals that Tutankhamun wasn't exactly a strapping sun god when he was alive.

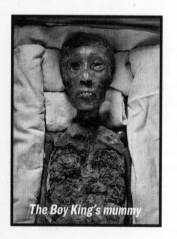

The Boy King's mummy

A new DNA study apparently solves several mysteries surrounding King Tut, including how he died and who his parents were. The study, published in the *Journal of the American Medical Association,* marks the first time the Egyptian government has allowed genetic studies to be performed using royal mummies. Analysis reveals that King Tut was a frail pharaoh, beset by malaria and a bone disorder, his health possibly compromised by newly discovered secrets of his parentage—Tut's mother and father were also brother and sister.

"Inbreeding is not an advantage for biological or genetic fitness. Normally the health and immune system are reduced and malformations increase," said study team member Carsten Pusch, a geneticist at Germany's University of Tübingen. "He was not a very strong pharaoh. He was not riding the chariots. Picture instead a frail, weak boy who had a bit of a club foot and who needed a cane to walk."

Life of Tut

Tutankhamun, pharaoh during ancient Egypt's New Kingdom era, about 3,300 years ago, ascended to the throne at the age of nine, but he ruled for only ten years before dying around 1324 B.C. Despite his brief reign, today King Tut is perhaps Egypt's best known pharaoh because of the wealth of treasure—including a solid gold death mask—found during the surprise discovery of his intact tomb in 1922.

Good Mummies, Good DNA

During the study, the condition of the DNA from the royal mummies of King Tut's family surprised many members of the team. Indeed, its quality was better than DNA gathered from nonroyal Egyptian mummies several centuries younger, Pusch said. The DNA of the Elder Lady (now identified as Queen Tiye, King Tut's paternal grandmother) for example, "was the most beautiful DNA that I've ever seen from an ancient specimen."

Art or Genetics?

King Tut's father, Akhenaten, is often depicted in artworks as having feminine features: wide hips, a potbelly, and femalelike breasts. Some speculated that he had a genetic disorder that caused him to develop these attributes, but when the team analyzed Akhenaten's body using medical scanners, no evidence of abnormalities were found. The team concluded that the feminized features found in the statues of Akhenaten created during his reign were done for religious, political, and artistic reasons.

The team suspects that the embalming method the ancient Egyptians used to preserve the royal mummies inadvertently protected DNA as well as flesh. "The ingredients used to embalm the royals was completely different in both quantity and quality compared to the normal population in ancient times," Pusch explained. Preserving DNA "was not the aim of the Egyptian priest of course, but the embalming method they used was lucky for us."

The Family Tree

In the new study, the mummies of King Tut and ten other royals that researchers have long suspected were his close relatives were examined. Of these ten, the identities of only three had been known for certain.

Using DNA samples taken from the mummies' bones, the scientists were able to create a five-generation family tree for the boy pharaoh. The team looked for shared genetic sequences in the Y chromosome—a bundle of

DNA passed only from father to son—to identify King Tut's male ancestors. The researchers then determined parentage for the mummies by looking for signs that a mummy's genes are a blend of a specific couple's DNA.

The Mummy's Daddy

In this way, the team was able to determine that a mummy known until now as KV55 is the "heretic king" Akhenaten—and that he was King Tut's father. Akhenaten was best known for abolishing ancient Egypt's pantheon in favor of worshipping only one god.

Furthermore, the mummy known as KV35 was King Tut's grandfather, the pharaoh Amenhotep III, whose reign was marked by unprecedented prosperity. Preliminary DNA evidence also indicates that two stillborn fetuses entombed with King Tut when he died were daughters whom he likely fathered with his chief queen Ankhensenamun, whose mummy may also have finally been identified.

> "What we can say for sure right now is that there is nothing wrong with [his] head. The head, is, indeed, intact."
>
> **Zahi Hawass**
> *Egyptian archaeologist, on whether or not Tut was killed by a blow to the head*

His Mother, His Aunt

King Tut's mother is a mummy researchers had been calling the Younger Lady.

While her body has finally been identified, her exact identity remains a mystery. DNA studies show that she was the daughter of Amenhotep III and Tiye and thus was the full sister of her husband, Akhenaten.

Some Egyptologists have speculated that King Tut's mother was Akhenaten's chief wife, Queen Nefertiti—made famous by an iconic bust. But the new findings seem to challenge this idea because historical records do not indicate that Nefertiti and Akhenaten were related.

Instead, the sister with whom Akhenaten fathered King Tut may have been a minor wife or concubine, which would not have been unusual, said Willeke Wendrich, a UCLA Egyptologist who was not involved in the study.

"Egyptian pharaohs had multiple wives, and often multiple sons who would potentially compete for the throne after the death of their father," Wendrich said.

Inbreeding would also not have been considered unusual among Egyptian royalty of the time.

His Left Foot

The team's examination of King Tut's body also revealed previously unknown deformations in the king's left foot, caused by the necrosis, or death, of bone tissue. "Necrosis is always bad, because it means you have dying organic matter inside your body," Pusch said. The affliction would have been painful and forced King Tut to walk with a cane—many of which were found in his tomb—but it would not have been life threatening.

What Killed the Boy King?

Malaria, however, would have been a serious danger. The scientists found DNA from the mosquito-borne parasite that causes malaria in the young pharaoh's body—the oldest known genetic proof of the disease.

The team found more than one strain of malaria parasite, indicating that King Tut caught multiple malarial infections during his life. The strains belong to the parasite responsible for malaria tropica, the most virulent and deadly form of the disease.

The malaria would have weakened King Tut's immune system and interfered with the healing of his foot. These factors, combined with the fracture in his left thighbone, which scientists had discovered in 2005, may have ultimately been what killed the young king, the authors write.

Until now the best guesses as to how King Tut died have included a hunting accident, a blood infection, a blow to the head, and poisoning. UCLA's Wendrich said the new finding "lays to rest the completely baseless theories about the murder of Tutankhamun." ∎

Tut's Teeth

The CT scans that revealed Tut's cause of death also showed that the king was well fed, stood 5 feet 6 inches (1.6 meters) tall, and had an overbite, much like the other kings in his family.

TRUTH:
KING TUT'S TOMB WAS FILLED WITH MORE THAN 5,000 OBJECTS.

"Vampire" Skull
Found in Italy

A skull unearthed in Italy reveals the very real belief in medieval vampires.

Near Venice, Italy, archaeologists found a woman's skull with a brick lodged in its jaws—an exorcism technique used on suspected vampires. It's the first time that archaeological remains have been interpreted as belonging to a suspected vampire, said team leader Matteo Borrini, a forensic archaeologist at the University of Florence. Borrini has been working on the island of Lazzaretto Nuovo, where the "vampire" was found, since 2006.

> "I was lucky. I [didn't] expect to find a vampire during my excavations."
> **Matteo Borrini**
> *University of Florence*

Bloody Belief

Belief in vampires was rampant mostly because decomposition was not well understood. For instance, as the human stomach decays, it releases a dark "purge fluid." This bloodlike liquid can flow freely from a corpse's nose and mouth, so it was apparently sometimes confused with traces of vampire victims' blood.

The fluid sometimes moistened the burial shroud near the corpse's mouth enough that it sagged into the jaw, creating tears in the cloth. Since tombs were often reopened during plagues so other victims could be added, Italian gravediggers saw these decomposing bodies with partially "eaten" shrouds.

Vampires were thought by some to cause plagues, so the superstition took root that shroud-chewing was the "magical way" that vampires spread pestilence, he said. Inserting objects—such as bricks and stones—into the mouths of alleged vampires was thought to halt the disease. ∎

In an ancient vampire-slaying ritual, a brick was lodged in the jaws of this 16th-century woman's skull.

Amelia Earhart

Spit Samples to Help Lick Mystery?

Amelia Earhart's dried spit could help solve the long-standing mystery of the aviator's 1937 disappearance, according to scientists who plan to harvest her DNA from envelopes.

Using Earhart's genes, a new project aims to create a genetic profile that could be used to test recent claims that her bones have been discovered. Right now, "Anyone can go and find a turtle shell and be like 'I found Amelia Earhart's remains,'" said Justin Long of Burnaby, Canada, whose family is partially funding the DNA project. The Internet-marketing executive is the grandson of 1970s aviator Elgen Long, who with his wife wrote the 1999 book *Amelia Earhart: The Mystery Solved*.

"We asked, How can we take wild claims like this and bring legitimacy back into the Amelia Earhart mystery?" Long said. "And so we started looking at everything at our disposal." According to Long, Earhart's letters are the only items that are both verifiably hers and that might contain her DNA. Hair would also be a good place to look for DNA, but no hair samples from Earhart are known. The International Woman's Air and Space Museum in Cleveland was once thought to have a lock of Earhart's hair, but a 2009 study revealed that the sample was actually thread.

KEY MOMENTS in the LIFE OF A LEGEND

| **1897:** Born in Atchison, Kansas | **1920:** Rides in a plane for the first time and decides to become a pilot. | **1921:** Completes flight school. Purchases her first aircraft. | **1922:** Sets women's world record for altitude by flying at 14,000 feet. | **1930:** Sets the women's world flying speed record of 181.18 mph. | **May 1932:** Becomes the first woman to fly solo across the Atlantic. |

Getting It on Paper

In July 1937, Earhart and her navigator, Fred Noonan, vanished over the central Pacific Ocean while attempting to fly around the world following the Equator. Earhart had already made history five years earlier, when she had become the first woman to fly solo across the Atlantic.

The remains of Earhart, Noonan, and their twin-engine plane were never recovered. But in 2009, researchers with the International Group for Historic Aircraft Recovery found a bone fragment on the South Pacific island of Nikumaroro, which they believed might have been from one of Earhart's fingers.

It should theoretically be possible, but no team has yet claimed to have extracted DNA from purported Earhart remains. Some scientists have even suggested the Nikumaroro bone fragment isn't human at all but may instead belong to a sea turtle whose remains were found nearby.

The new Earhart DNA project will be headed by Dongya Yang, a genetic archaeologist at Simon Fraser University in Burnaby, Canada. Yang will examine Earhart's letters and attempt to extract DNA from the saliva she used to seal the envelopes.

Mining a trove of more than 400 correspondences between Earhart and various people, the researchers have chosen four letters to family—deemed the most likely to have been written and sealed by Earhart herself—for analysis.

Earhart "did have her own secretary, so it's most likely that anything that was business related was done by the secretary," project funder Long said. "But if she's just at home writing a personal letter, there's much less reason for the secretary to be involved."

How Did Earhart Lick?

Fortunately for the team, people in Earhart's time typically opened letters from the side, using a letter opener, so the original seals haven't been broken. Yang is aiming to gather two kinds of DNA from the letters: mitochondrial DNA, which children inherit from their mothers only, and nuclear DNA,

August 1932:	1935:	March 1937:	June 1937:	July 2, 1937:	July 19, 1937:
Becomes the first woman to fly nonstop across the U.S.	Becomes the first woman to fly solo across the Pacific Ocean.	Begins a round-the-world flight, but abandons the attempt because of damage to her plane.	Starts a second round-the-world trip from Miami, Florida, with Fred Noonan,	After departing New Guinea, her plane disappears over the Pacific.	Official search efforts end.

which contains the bulk of a person's genetic information and is housed in each human cell's nucleus. If both DNA types can be obtained, the team says it can create a genetic profile of Earhart that is complete enough to positively identify any potential remains.

The DNA-harvesting technique is thought to be nondestructive, but just in case, Yang is perfecting his procedure on test envelopes first. "When we have the best technique available, that's when we'll move on to the real letters," Yang said.

Yang thinks it's very likely his team will find cells on the envelope seals, but how many cells are present will depend in part on how Earhart sealed her letters. "A strong licking or a light licking may leave more or less cells with the seal," Yang said.

> "For food I carried a very simple ration— tomato juice. I think that serves as food and drink, and I used just a few swallows of it . . . The fact is, one doesn't think much about food on such a journey."
> **Amelia Earhart**
> accepting the National Geographic Society's Special Medal for being the first woman to make a solo transatlantic crossing, 1932

Wanted: DNA

Geneticist Brenna Henn said she knows of no other case where DNA has been harvested from decades old letters. But she said Yang's methodology "sounds reasonable." The team will need "quite a bit of DNA" to succeed, though.

Extracting nuclear DNA will be challenging, since each cell has only one copy, said Henn, of Stanford University. "You need more than just one copy of the cell in order for the [nuclear DNA] amplification to work," said Henn, who's not involved in the Earhart project.

Also, if the team obtains only short fragments of Earhart's nuclear DNA, the genes may not be useful for identification purposes, she added. "The problem is the majority"—about 99 percent—"of the nuclear genome is identical among all humans," she said. "If they could obtain little fragments, they have almost no power to discriminate between Earhart's DNA and that of other living people."

To ensure that the DNA from the letters indeed belonged to Earhart, the team will compare it to DNA from Earhart's still-living relatives and also DNA extracted from another letter, written by Earhart's sister and addressed to Elgen Long. If the project proceeds smoothly, Yang said, the team could have a genetic profile for Earhart soon. ∎

Great Pyramid Mystery

To Be Solved by Hidden Room?

A sealed space in Egypt's Great Pyramid may help solve a centuries-old mystery: How did the ancient Egyptians move two million 2.5-ton blocks to build the ancient wonder?

A little-known cavity in the 4,500-year-old monument to Pharaoh Khufu may support the theory that the pyramid was constructed inside out, via a spiraling, inclined interior tunnel—an idea that contradicts the prevailing wisdom that the monuments were built using an external ramp. The inside-out theory's key proponent, French architect Jean-Pierre Houdin, says for centuries Egyptologists have ignored evidence staring them in the face. "The paradigm was wrong," Houdin said. "The idea that the pyramids were built from the outside was just wrong. How can you resolve a problem when the first element you introduce in your thinking is wrong?"

> ### Stone Cold Facts
>
> 1. The heaviest blocks are estimated to weigh as much as 70 tons.
> 2. There are no hieroglyphics or writing on the stones in the Great Pyramid.
> 3. Two types of limestone were used for construction: a soft limestone for the core blocks and a hard white limestone for the outside.

Finding Flaws

Even the most widely held Great Pyramid construction theories have flaws, Egyptologist Bob Brier said. For example, a single, straight external ramp would have been impractical, said Brier, of Long Island University in New York.

To deliver blocks to the 481-foot (147-meter) peak at a reasonable grade, the ramp would have had to have been a mile (1.6 kilometers) long and made of stone. And during the decades of the pyramid's construction, workers would have had to continually increase the ramp's height and length as the pyramid rose. "That's like building two pyramids. And we've never found the remains of such a ramp," Brier said.

Another theory suggests a stone ramp wound around the outside of the Great Pyramid. But an outside ramp would have obscured the pyramid's surface—making it impossible for surveyors to use the corners and edges for necessary calculations during construction, Brier said.

Greek historian Herodotus, writing around 450 B.C., theorized the use of small, wooden cranes or levers to lift the blocks. But, Brier said, "you'd have to have thousands, and they didn't have enough wood in all of Egypt for that," Brier said.

Like Father, Like Son

For Houdin, the Paris architect, the puzzle of the pyramid is a family affair. His father, a civil engineer, came up with the idea of an internal construction ramp a decade ago. Houdin was soon hooked, as suggested by his books, co-written by Brier—*The Secret of the Great Pyramid: How One Man's Obsession Led to the Solution of Ancient Egypt's Greatest Mystery.* Houdin eventually left his architecture firm to pursue the inside-out theory full-time.

For what they thought would be a matter of weeks, he and his wife moved into a 236-square-foot (22-square-meter) studio apartment. They ended up staying for four years, as Houdin toiled away at his self-financed project.

From the Inside Out

Houdin's theory suggests the Great Pyramid was built in two stages. First, blocks were hauled up a straight external ramp to build the pyramid's bottom third, which contains most of the monument's mass, Houdin believes. Houdin says the limestone blocks used in the outside ramp were recycled for the pyramid's upper levels, which might explain why no trace of an original ramp has been found.

Egyptian-archaeology specialist Josef Wegner sees merit in the recycling idea. "The notion of using the already quarried smaller blocks to build the lower ramp and then dismantling that for use in upper sections would be a very logical approach to speed up the overall construction process," said Wegner of University of Pennsylvania Museum of Archaeology and Anthropology.

After the foundation had been finished, workers began building an inclined, internal, corkscrew tunnel, which would continue its path up and around as the pyramid rose, Houdin said. Because the tunnel is inside the pyramid, Brier said, "when they finished getting blocks all the way up to the top this ramp disappeared [from view]."

The Secret of the Hidden Room

New evidence uncovered about two-thirds of the way up the Great Pyramid supports the inside-out theory, said Houdin. At about the 300-foot (90-meter) mark on the northeastern edge lies an open notch.

On a recent expedition with a National Geographic film crew, Brier—aided by a videographer with mountain-climbing experience—scaled perilous crumbling rocks to reach the notch. Ducking inside the notch, Brier entered a small L-shaped room. He wasn't the first to visit the space, but until now Egyptologists had taken little notice of it. Houdin said the feature figures perfectly with his theory.

Open Corners for Turning Blocks?

For the interior tunnel to work, it would have required open areas at the Great Pyramid's four corners, Houdin says. Otherwise the blocks wouldn't have been able to clear the 90-degree turns. Like railroad roundhouses, these open corners would have given workers room to pivot the blocks—perhaps using wooden cranes—so the stones could be pushed into the next tunnel.

The notch and room are remnants of one such opening, Houdin claims. They are located at one of the spots where Houdin's 3-D computer models suggest they should be. Inside the corner space, which was apparently walled in as the pyramid was completed, there should be two tunnel entrances at right angles to each other—each leading to a section of the internal ramp, Houdin believes. Perhaps all that stands between him and the solution to the mystery are massive blocks that thousands of years ago sealed the tunnel, Houdin said.

The Great Pyramid of Khufu

If this previously known space truly is the missing link in the puzzle of the Great Pyramid's construction, the question remains why no one would have surmised this by now. Brier said, "If you weren't thinking about internal ramps and notches and you climbed right by this thing, it wouldn't mean anything to you."

An Important Clue

Prior to the room brainstorm, Houdin's most important piece of evidence was the product of good luck. In 1986, a French team in an ultimately fruitless search for hidden chambers in the Great Pyramid had done a survey of the monument's density using a technique called microgravimetry, which measures the strength of local gravitational fields.

Nearly 15 years later, Houdin was presenting his ramp theory at a conference and was approached by a member of the 1986 team. The man showed Houdin an image from their survey that they'd dismissed as unexplainable. But to Houdin, and later Brier, the explanation was clear. The image shows what looks like a spiraling feature inside the structure's outer walls. "If I hadn't seen that diagram, I'd probably be thinking this is just another theory," Brier said.

Seeking Confirmation

The 1986 image, the notch room, and other evidence may make Houdin's theory plausible, but the case is far from closed. "As with all archaeological theories, the proof is in the pudding, and many logical and compelling theories have fallen by the wayside under the weight of hard evidence," said the University of Pennsylvania's Wegner. But "verification of the proposed internal spiral ramp would be a remarkable and groundbreaking discovery," Wegner added.

Houdin believes that verification is possible. He has suggested that an infrared camera—positioned about 150 feet (46 meters) from the pyramid—could potentially record subtle differences in interior materials and temperatures. Those variations could reveal clear-cut "phantoms" of the internal ramp.

"What we need is the authorization, by the Egyptian authorities, to stay around for 18 hours, close to the pyramid, with a cooled infrared camera based on an SUV and to take images of three [pyramid] faces every hour during this period," Houdin said. "A green light from Cairo and the Great Pyramid mystery could be over." ■

Built for a King

Egyptologists believe that the pyramid was built as a tomb for fourth-dynasty Egyptian pharaoh Khufu, who ruled from 2589 to 2566 B.C. Not only was Khufu honored by the pyramid, he was also its architect. The giant structure took more than 20 years to build, and construction finished around 2560 B.C. For almost 4,000 years, it was the tallest man-made structure on Earth.

The Freemasons
8 Myths Decoded

Books like Dan Brown's *The Lost Symbol* have shined a light on secret societies like the Freemasons, generating new interest and new conspiracy theories about them. But what's true?

Society's Secrets: Freemasons have been accused of everything from conspiring with extraterrestrials to practicing sexual deviancy to engaging in occult rituals to running the world—or trying to end it. Detractors include global conspiracy theorists and religious organizations, including the Catholic Church.

But what if Freemasons—the world's largest international secret society—are just a bunch of guys into socializing, nonsatanic rituals, self-improvement, and community service?

To separate Freemason fact from myth, National Geographic News went inside the centuries-old order with two Masons and a historian of the ancient Christian order from which some claim the Masons sprang in the 17th or 18th century.

MYTH 1 Masonic Symbols Are Everywhere

It's true that Masonic symbols are anything but lost, said Freemason and historian Jay Kinney, author of the newly released *Masonic Myth*.

Freemasonry is rich in symbols, and many are ubiquitous—think of the pentagram, or five-pointed star, or the "all-seeing eye" in the Great Seal of the United States. But most Masonic symbols aren't unique to Freemasonry, Kinney said.

"I view the Masonic use of symbols as a grab bag taken from here, there, and everywhere," he said. "Masonry employs them in its own fashion." The pentagram, for example, is much older than Freemasonry and acquired its occult overtones only in the 19th and 20th centuries, hundreds of years after the Masons had adopted the symbol.

Likewise, the all-seeing eye saw its way to the Great Seal—and the U.S. dollar bill—by way of artist Pierre Du Simitiere, a non-Mason. The eye represents divine guidance of the U.S. ship of state, or as Secretary of the U.S. Congress Charles Thompson put it in 1782, it alludes "to the many signal interpositions of providence in favour of the American cause." There was one known Mason on the committee to design the seal: Benjamin Franklin. His proposed design was eyeless, and rejected.

MYTH 2 Masons Descend From the Knights Templar

Much has been made of the Freemasons purported lineage to the Knights Templar. The powerful military and religious order was established to protect medieval pilgrims to the Holy Land and dissolved by Pope Clement V, under pressure of King Phillip IV of France, in 1312.

After modern Masonry appeared in 17th- or 18th-century Britain, some Freemasons claimed to have acquired the secrets of the Templars and adopted Templar symbols and terminology—naming certain levels of Masonic hierarchy after Templar "degrees," for example. "But those [Knights Templar] degrees and Masonic orders had no historic connection with the original Knights Templar," Kinney explained.

"These are myths or symbolic figures that were used by the Masons. But because the association had been made with these degrees, and the degrees had perpetuated themselves, after a time it began to look like there had been a connection."

Helen Nicholson, author of *The Knights Templar: A New History*, agrees that there is no possibility that Freemasons are somehow descended from the Knights Templar. By the time of the first Masons, the Cardiff University historian said, "there were no more Templars."

MYTH 3 Masons Are Hiding Templar Treasure

One of the Templar-Mason theory's many veins suggests that some Templars survived the order's 14th-century destruction by taking refuge in Scotland, where they hid a fabulous treasure beneath Rosslyn Chapel. The treasure, and the Templar tradition, were eventually passed down to the founders of Freemasonry, the story goes.

In fact, there was Templar treasure, Nicholson said, but it ended up in other hands long ago. "The most likely reason [the Templars were dissolved] is that the king wanted their money. The King of France was bankrupt, and the Templars had lots of ready cash."

MYTH 4 Washington, D.C.'s Streets Form Giant Masonic Symbols

It's long been suggested that powerful Freemasons embedded Masonic symbols in the Washington, D.C., street plan designed mainly by Frenchman Pierre L'Enfant in 1791.

"Individually, Masons had a role in building the White House, in building and designing Washington, D.C.," said Mark Tabbert, director of collections at the George Washington Masonic Memorial in Alexandria, Virginia. "And [small scale] Masonic symbols can be found throughout the city, as they can in most U.S. cities."

But there's no Masonic message in the city's street plan, Tabbert said. For starters, Pierre L'Enfant wasn't a Mason. And, Tabbert asked, why would Masons go to the trouble of laying out a street grid to match their symbols? "There has to be a [reason] for doing such a thing," said Tabbert, himself a Mason. "But there isn't one."

MYTH 5 Freemasons Rule the World

Maybe it's the impressive list of prominent Freemasons—from Napoleon to F.D.R. to King Kamehameha (IV and V)—that's led some to suggest the group is a small cabal running the globe. But Kinney, the Masonic historian, paints a picture of a largely decentralized group that might have trouble running anything with much efficiency.

"I think the ideals that Masonry embodies, which have to do with universal brotherhood, are shared by Masons around the world [regardless of] religious, political, or national differences," he said. "But having shared ideals is one thing—having some sort of shared hierarchy is something else altogether."

Kinney noted that the United States alone has 51 grand lodges, one for each state and the District of Columbia. Each of these largely independent organizations oversees its many local blue (or beginner) lodges and has little real coordination with other grand lodges.

Internationally, Masonic lodges not only don't speak with a single voice but sometimes refuse to even recognize one another's existence. Also, many Masons are independent minded and tend to resist edicts from above, Kinney said. "There is no way that they could be run by a single hierarchy. There is no such entity."

MYTH 6   Freemasonry Is a Religion—Or a Cult

Masons stress that their organization is not a religion—that is, it has no unique theology and does not represent a path for believers to salvation or other divine rewards. Even so, to be accepted into Freemasonry, initiates must believe in a god—any god. Christians may be in the majority, but Jews, Muslims, and others are well represented in Masonic circles. At lodge meetings religious discussion is traditionally taboo, Kinney and Tabbert said.

MYTH 7   Freemasons Started the American Revolution

Prominent Freemasons like Ben Franklin and George Washington played essential roles in the American Revolution. And among the ranks of Freemasons are nine signers of the Declaration of Independence and 13 signers of the Constitution. But Freemasonry—born in Britain, after all—had adherents on both sides of the conflict. Tabbert, of the George Washington Masonic Memorial, said Masonic groups allowed men on both sides of the Revolution to come together as brothers—not to promote a political view, which would be against Masonic tradition.

"For many years [Masons] claimed in their own quasi-scholarship that all of these revolutionaries and Founding Fathers were Freemasons," Tabbert said. "A fair number of them were, but they weren't doing these things because they were Freemasons."

MYTH 8   Membership Requires Shadowy Connections

Contrary to popular fiction, you don't have to drink wine from a skull to become a ranking Freemason. In fact, tradition dictates that Masons don't recruit members but simply accept those who approach them of their own free will.

When Freemasonry hit its peak in the United States during the late 1950s, Kinney said, almost one of every ten eligible adult males was a member—a total of some four million and hardly a tiny elite.

Today membership numbers, like those of other fraternal organizations, have declined dramatically, and only about 1.5 million U.S. men are Masons.

But with perhaps new interest generated by pop culture and conspiracy theories, Masonic centers should brace for tourists—and maybe a few new recruits. ∎

Crop Circles Explained!
Frauds, Artists, Scientists, and Aliens

Are crop circles—flattened, gigantic patterns in farmers' fields—agrarian graffiti, large-scale land art, or something more profound . . . an otherworldly message from outer space?

Crop circles create controversy. Many believe that they are the work of artists—people who steal into fields of wheat and barley to flatten stalks into intricate and otherworldly patterns. Others believe they are of supernatural origin. No matter what side you're on, the debate fascinates and draws tens of thousands of people to the English countryside every year to have a look for themselves.

English Roots

Crop circles began to appear in the fields of southern England in the mid-1970s. Early circles were quite simple, and simply appeared, overnight, in fields of wheat, rape, oat, and barley. The crops are flattened, the stalks bent but not broken.

What's My Name?

Crop-circle enthusiasts, or "croppies," also call themselves cereologists—after Ceres, the Roman goddess of agriculture. Most believe that crop circles are the work of either extraterrestrials or plasma vortices.

Wiltshire County is the acknowledged center of the phenomenon. The county is home to some of the most sacred Neolithic sites in Europe, built as far back as 4,600 years ago, including Stonehenge, Avebury, Silbury Hill, and burial grounds such as West Kennet Long Barrow.

As the crop circle phenomenon gained momentum, formations

Giant crop circles in a wheat field in Wiltshire, England

have also been reported in Australia, South Africa, China, Russia, and many other countries, frequently in close proximity to ancient sacred sites. Still, each year more than a hundred formations appear in the fields of southern England.

Artists' Creations

In 1991, two artists, Doug Bower and Dave Chorley, came forward and claimed responsibility for crop circles appearing in the '70s and '80s. "I think Doug Bower is the greatest artist of the 20th century," said John Lundberg, a graphic design artist, website creator, and acknowledged circle maker. Bower's work has the earmarks of all new art forms, "pushing boundaries, opening new doors, working outside of the established mediums," Lundberg continued. Lundberg works with a group, known as the Circlemakers, a dedicated crop circle art group in the United Kingdom. Circlemakers now engages in quite a bit of commercial work; the group has created a giant crop formation 140 feet (46 meters) in diameter for the History Channel. But they also still do covert work in the dead of night.

> **TRUTH:**
> THE EARLIEST RECORDED MENTION OF A CROP CIRCLE DATES BACK TO THE 1500S.

Circle Time

Formulating a design and a plan, from original concept to finished product, can take up to a week. "It has to be more than a pretty picture. You have to have construction diagrams providing the measurements, marking the center, and so on," said Lundberg. Creating the art is the work of a night.

Lundberg said that for an artist, being a crop-formation artist is an interesting place to be, but circle makers rarely claim credit for specific formations they created. "To do so would drain the mystery of crop circles," he explained. "The art form isn't just about the pattern making. The myths and folklore and energy [that] people give them are part of the art."

Crop Circles of Note

1. 2005, Hampshire, United Kingdom: A giant rendering of an alien from the classic arcade game Space Invaders.
2. 2008, Wiltshire, United Kingdom: 150-foot-wide (46-meter) coded representation of the first ten digits of the mathematical constant pi appeared.
3. 2009 Oxfordshire, United Kingdom: A 600-foot-long (182-meter) jellyfish almost engulfed a barley field.
4. 2009, Cyberspace: The Google Doodle turned into a series of crop circles, complete with a circling UFO.

Increasing Complexity

During the last 25 years, the formations have evolved from simple, relatively small circles to huge designs with multiple circles, elaborate pictograms, and shapes that invoke complex nonlinear mathematical principles. A formation that appeared in August 2001 at Milk Hill in Wiltshire contained 409 circles, covered about 12 acres (5 hectares), and was more than 800 feet (243 meters) across.

To combat the theory that the circles were the result of wind vortices—essentially mini-whirlwinds—crop artists felt compelled to produce ever more elaborate designs, some with straight lines to show that the circles were not a natural phenomenon, said Lundberg.

On the question of whether all such circles are human made, Lundberg is perched firmly on the fence. "I don't care," he said. "I have an open mind. It would be great if people could view circles as an art form. But really, to me, as long as they're well made and well crafted, anyone can believe whatever they want to believe."

Unknown Origins

There are groups who strongly oppose the artists. These researchers of the paranormal and scientists seek to explain the formations as work that could not possibly be the result of human efforts. Some believers are merely curious, open to the existence of paranormal activity and willing to consider the possibility that at least some of the circles were created by extraterrestrial forces.

U.K. crop circle researcher Karen Alexander believes there is room for both schools of thought. "There's no doubt that some crop circles are made by people and that some are made for advertising campaigns and so forth," said Alexander, co-author of *Crop Circles: Signs, Wonders, and Mysteries.* "But then there's quite a large percentage for which origin is just unknown."

Alternative explanations range from natural phenomena such as whirlwinds to visitations by aliens in UFOs. "There are lots of other equally strange ideas," Alexander said. "People talk about earth energies, or think that earth spirits perhaps make them."

"Crop circles 'like' ancient sites," she noted. Wiltshire—generally acknowledged as crop circle central—is littered with Stone Age monuments, including Stonehenge. Alexander herself remains "totally openminded." "I just think they're a fascinating cultural phenomenon," she said.

A Mighty Wind

One scientific explanation given for crop circles is that they are akin to dust devils and created by small currents of swirling winds. The spinning columns force a burst of air down to the ground, which flattens the crops. Dr. Terence Meaden of the Tornado and Storm Research Organisation (TORRO) in Wiltshire, England, says the vortices that create crop circles are charged with energy. When dust particles get caught up in the spinning, charged air, they can appear to glow, which may explain the UFO-like glowing lights many witnesses have seen near crop circles.

Circle Gawkers

Crop circle season extends from roughly April to harvesting in September—and the Wiltshire community profits from it. The best time to make (and to see) a circle is in mid to late June, making summer crop circles popular tourist attractions in some corners. They've spawned bus tours, daily helicopter tours, and sales of T-shirts, books, and other trinkets. "We get tens of thousands of people coming to the U.K. each year just to look at them," Alexander said. And the phenomenon shows no signs of abating. "I think we've had more crop circles in the U.K. than just about any year I can remember," Alexander said. ∎

7 Moon Landing Hoax Myths—Busted

You don't have to be a rocket scientist to overturn these myths (although it wouldn't hurt).

Forty years after U.S. astronaut Neil Armstrong became the first human to set foot on the moon, many conspiracy theorists still insist the *Apollo 11* moon landing was an elaborate hoax. Examine the evidence, and find out why experts say some of the most common claims simply don't hold water.

MYTH 1 The Flags Were Waving

You can tell the moon landing was faked because the American flag appears to be flapping as if "in a breeze" in videos and photographs supposedly taken from the airless lunar surface.

While on the moon, the astronauts accidentally bent the horizontal rods holding the flag in place several times, creating the appearance of a rippling flag in photographs.

The fact of the matter is "the video you see where the flag's moving is because the astronaut just placed it there, and the inertia from when they let go kept it moving," said spaceflight historian Roger Launius, of the Smithsonian's National Air and Space Museum in Washington, D.C.

MYTH 2 Mystery Photographer

You can tell the moon landing was faked because only two astronauts walked on the moon at a time, yet in photographs such as this one where both are visible, there is no sign of a camera. So who took the picture?

The fact of the matter is the cameras were mounted to the astronauts' chests, said astronomer Phil Plait, author of the award-winning blog Bad Astronomy and president of the James Randi Educational Foundation. In the picture below, Plait notes, "you can see [Neil's] arms are sort of at his chest. That's where the camera is. He wasn't holding it up to his visor."

Neil Armstrong and the Eagle lunar lander are reflected in Buzz Aldrin's visor in one of the most famous images taken during the July 1969 moon landing.

MYTH 3 Where Are the Stars?

You can tell the moon landing was faked because the astronauts made no such exclamation while on the moon, and the black backgrounds of their photographs are curiously devoid of stars.

A view of the moon's star-free sky as seen from the surface.

The fact of the matter is the moon's surface reflects sunlight, and that glare would have made stars difficult to see. Also, the astronauts photographed their lunar adventures using fast exposure settings, which would have limited incoming background light. "They were taking pictures at 1/150th or 1/250th of a second," Bad Astronomy's Plait said. "In that amount of time, stars just don't show up."

MYTH 4 — No Landing Crater

You can tell the moon landing was faked because the module is shown sitting on relatively flat, undisturbed soil. According to skeptics, the lander's descent should have been accompanied by a large dust cloud and would have formed a noticeable crater.

The lunar lander known as the Eagle rests peacefully on the moon's surface in a picture taken mere hours after the July 20, 1969, moon landing.

The fact of the matter is the lander's engines were throttled back just before landing, and it did not hover long enough to form a crater or kick up much dust, the Smithsonian's Launius said. "Science fiction movies depict this big jet of fire coming out as [spacecraft] land, but that's not how they did it on the moon," he added. "That's not the way they would do it now or anytime in the future."

MYTH 5	## Strange Shadows

You can tell the moon landing was faked because Aldrin is seen in the shadow of the lander, yet he is clearly visible. Hoax subscribers say that many shadows look strange in Apollo pictures. Some shadows don't appear to be parallel with each other, and some objects in shadow appear well lit, hinting that light was coming from multiple sources—suspiciously like studio cameras.

The fact of the matter is there were multiple light sources, Launius said. "You've got the sun, the Earth's reflected light, light reflecting off the lunar module, the space suits, and also the lunar surface." It's also important to note that the lunar surface is not flat, he added. "If an object is in a dip, you're going to get a different shadow compared to an object next to it that is on a level surface."

A moon-landing picture shows astronaut Buzz Aldrin standing on the footpad of the Eagle's ladder, his bent knees suggesting that he's about to jump up to the next rung.

MYTH 6 **Footprints Too Clear**

You can tell the moon landing was faked because the astronauts' prints are a bit too clear for being made on a bone-dry world. Prints that well defined could only have been made in wet sand.

The fact of the matter is that's nonsense, said Bad Astronomy's Plait. Moon dust, or regolith, is "like a finely ground powder. When you look at it under a microscope, it almost looks like volcanic ash. So when you step on it, it can compress very easily into the shape of a boot." And those shapes could stay pristine for a long while thanks to the airless vacuum on the moon.

The contrasted lines of a boot print appear as Buzz Aldrin lifts his foot to record an image for studying the moon's soil properties. Apollo pictures show scores of clear boot prints left behind as the astronauts traipsed across the moon.

MYTH 7 **No Leftovers**

You can tell the moon landing was faked because with instruments such as the Hubble Space Telescope capable of peering into the distant recesses of the universe, surely scientists should be able to see the various objects still on the moon. But no such pictures of these objects exist.

The fact of the matter is no telescope on Earth or in space has that kind of resolving power. "You can calculate this," Plait said. "Even with the biggest telescope on Earth, the smallest thing you can see on the surface of the moon is something bigger than a house." ■

When Armstrong and Aldrin took off from the moon in July 1969, they left behind part of the Eagle, the U.S. flag, and several other instruments and mementos, including the seismometer Aldrin is adjusting in the above picture.

Loch Ness Sea Monster

Fossil a Fake, Say Scientists

When the remains of a giant sea creature were discovered in Scotland's Loch Ness, the world went wild. But why were scientists so skeptical? Is it all just a hoax?

Along the shores of Loch Ness, Scotland, the discovery of fossilized remains surprised and delighted fans of the world's most famous monster. Could they belong to the Loch Ness Monster? Many people believe "Nessie" to be alive and well, if rather

Gerald McSorely holds up the fossil he found on the shores of Loch Ness.

shy, having sought refuge in Scotland's largest freshwater loch. Of course, there are still plenty of skeptics who believe there is no Nessie—alive or dead. Could these four fossilized vertebrae, complete with blood vessels and spinal column, be the sort of thing that could convince them?

First News of Nessie

The Loch Ness Monster legend is said to date back more than 1,400 years, when Saint Columba encountered a strange water beast in the region. But it kept a low profile until 1933, when a new road made the loch more accessible and gave clear views from its northern shore. A flood of reported sightings soon followed, the first coming from an innkeeper at Drumnadrochit.

That same year saw the publication of the most famous picture of Nessie—its neck and head rising from the loch's murky waters. Taken by a respected gynecologist, Colonel Robert Wilson, the monster became an overnight sensation.

But in 1994 Wilson's photograph made the front pages again—when exposed as one of the greatest hoaxes of the 20th century. Christian Spurling confessed shortly before his death that the grainy, black-and-white image actually showed a piece of plastic attached to a toy submarine. Spurling made the model for his stepfather, Marmaduke Wetherall, who, along with Wilson, wanted something to show for their monster hunting expedition.

It seems the hoaxers have been trying to match their success ever since.

Dem Bones

But were the bones found in 2003 part of a hoax? Retired scrap metal dealer Gerald McSorley, from Stirling in Scotland, found the remains. The pensioner said he chanced upon it when he tripped and fell in the loch. He told the world: "I have always believed in the Loch Ness Monster, but this proves it for me. The resemblance between this and the sightings which have been made are so similar."

> **TRUTH:**
> THE LIFE OF GARY CAMPBELL, PRESIDENT OF THE OFFICIAL LOCH NESS MONSTER FAN CLUB, IS INSURED AGAINST HIM BEING EATEN ALIVE BY THE MONSTER, TO THE SUM OF £250,000 (U.S. $400,000).

His discovery was confirmed by staff at the National Museum of Scotland in Edinburgh. Paleontologists determined that the fossilized bones did belong to a plesiosaur, a fearsome predator that ruled the seas between 200 and 65 million years ago. Measuring 35 feet (11 meters) head to tail, with a long, serpentine neck, the reptile eventually died out with the dinosaurs. Yet they still had their doubts.

Planted Evidence?

Lyall Anderson, one of the museum's paleontologists, said: "The fossil is definitely that of a plesiosaur—a very good example. And I believe Mr. McSorley when he says he found it where he did. But there's evidence to suggest it came from elsewhere and had been planted.

"The fossil is embedded in a gray, Jurassic-aged limestone. Rocks in the Loch Ness area are much older—they're all crystalline, igneous, and metamorphic rocks." Anderson says the nearest match for this limestone is at Eathie on the Black Isle, some 30 miles (50 kilometers) northeast of Loch Ness. He added: "The stone has been intensely drilled by marine organisms. It seems likely the specimen was on a seashore until relatively recently."

Sea Monster

Others agree with Anderson. Richard Forrest, a plesiosaurus expert at the New Walk Museum in Leicester, England, said: "The fossil's general appearance, and the presence of holes made by burrowing sponges, shows it has spent some time in the sea, probably [with] beach pebble[s]. Yet Loch Ness contains freshwater."

Gary Campbell, president of the Official Loch Ness Monster Fan Club, added: "I think it's almost certain the fossil was placed there deliberately.

Other Lakes, Other Monsters

Loch Ness isn't the only place that's home to a legendary lake creature. These other places around the world have monsters of their own:

1. "Wally" Wallowa Lake, Oregon, U.S.A.
2. "Ogopogo" Okanagan Lake, British Columbia, Canada
3. "Manipogo" Lake Manitoba, Manitoba, Canada
4. "Cressie" Crescent Lake, Newfoundland, Canada
5. "Hamlet" Lade Elsinore, California, U.S.A.
6. "Tahoe Tessie" Lake Tahoe, California, U.S.A.
7. "Champ" Lake Champlain, New York/Vermont, U.S.A.
8. "Auli" Lake Chad, Chad, Africa
9. "Chipekwe" Lake Tanganyika, Tanzania, Africa
10. "Morag" Loch Morar, Scotland, U.K.
11. "Brosno Dragon" Brosno Lake, Andreapol, Russia
12. "Issie" Lake Ikeda, Kagoshima, Japan

There are very few public access points to the shore and the fossil was found by a layby where lots of tourists stop. As far as we're concerned it was left for someone to find."

Campbell says numerous other Nessie-inspired hoaxes have taken place in this area. Recently conger eels were dumped in the loch after being caught by sea fishermen, presumably in the hope they would be mistaken for miniature monsters. One measured more than 6 feet (2 meters) long.

However, Forrest says there could also be an innocent reason why the fossil turned up here. He said: "A plesiosaur limb bone was found near the same spot some years ago. It turned out it belonged to a local tour operator, who used the fossil as a demonstration piece. He'd left it on a rock and forgot about it."

Probably Not a Plesiosaur

With Nessie's true likeness still shrouded in mystery, Anderson says the plesiosaur is the creature that most commonly springs to mind, adding: "It's an image many people are familiar with and they often try to pin it on the Loch Ness Monster." But paleontologists say this is wishful thinking. While plesiosaurs died out many millions of years ago, Loch Ness is less than 12,000 years old, having been glacially excavated during the last ice age.

Forrest also adds that plesiosaurs, being cold-blooded reptiles, wouldn't generate enough internal body heat to survive the loch's cool temperatures. And even if they could, there wouldn't be enough food for them to survive.

Plus, plesiosaurs breathed air and would need to surface several times a day, at the very least. "Despite this, I haven't heard of any sightings that sound anything like a plesiosaur," Forrest said. "People usually refer to a series of undulating humps. The plesiosaur, being a reptile, wouldn't undulate but move from side to side. Such sightings are much more likely to come from mammals—such as a row of otters swimming across the loch."

Forrest concludes: "My own view is that reports of the Loch Ness Monster are very good for the Scottish tourist industry, but not backed up by any real evidence. One thing is for sure: Even if there is a large animal in Loch Ness, it's not a plesiosaur." ■

CHAPTER 6

Feath
Friends

ered

There's a lot more to birds than just colorful feathers and beautiful songs. Beneath their ordinary exteriors lurk creatures that are unexpectedly weird, like the half-male, half-female chicken and the bird that sings through its feathers. There are smiling birds and smelly birds. There are birds that fly as high as airplanes and others that are as intelligent as humans. So don't underestimate the birds; they're a whole lot weirder than meets the eye.

Highest Flying Bird Found

The bar-headed goose can reach nearly 21,120 feet and fly over the Himalaya in just about eight hours, a new study shows.

The world's highest flying bird is an Asian goose that can fly up and over the Himalaya in only about eight hours, a new study finds. The bar-headed goose is "very pretty, but I guess it doesn't look like a super-athlete," said study co-author Lucy Hawkes, a biologist at Bangor University in the United Kingdom.

A Little Traveling Music, Please

In 2009, Hawkes and an international team of researchers tagged 25 bar-headed geese in India with GPS transmitters. Shortly thereafter, the birds left on their annual spring migration to Mongolia and surrounding areas to breed.

To get there, the geese have to fly over the Himalaya—the world's tallest mountain range and home to the tallest mountain on Earth, Mount Everest, which rises to 29,035 feet (8,850 meters). The researchers found that the birds reached a peak height of nearly 21,120 feet (6,437 meters) during their travels. The migration took about two months and covered distances as long as 5,000 miles (8,000 kilometers).

The birds made frequent rest stops during the migration, but they appear to have flown over the Himalayan portion of their journey in a single effort that took about eight hours on average and included little or no rest. A similar intense climb could kill a human without proper acclimatization, Hawkes said.

"If you've ever seen a goose sitting on a lake, take-off is quite an energetic thing, so it may be [energetically] cheaper to keep going than to keep sitting down and taking off again—and they may not want to delay getting over the mountains," Hawkes said.

A Bod Built for Flight

Even more impressive, the birds completed the ascent under their own muscular power, with almost no aid from tail winds or updrafts. "Most other species that we've identified as high-altitude flyers usually get there by soaring and gliding up," said study leader Charles Bishop, also a biologist at Bangor University.

TRUTH: BAR-HEADED GEESE HAVE A SPECIAL TYPE OF HEMOGLOBIN THAT ABSORBS OXYGEN MORE QUICKLY THAN THAT OF OTHER BIRDS, SO THEY CAN EXTRACT MORE OXYGEN FROM EACH BREATH.

By contrast, the bar-headed goose reaches such lofty heights by flapping vigorously, if not gracefully. "Geese tend to honk a lot as they fly," co-author Hawkes said. "We don't think of them as the most elegant of migrants."

The birds have evolved numerous physiological adaptations—many of which are not so obvious—to help them complete their migrations. "They

Bar-headed geese can fly over the Himalaya in eight hours.

have all these internal morphological tweaks that make it possible," Bishop said. For example, past studies have shown that the geese have more capillaries and more efficient red blood cells than other birds, meaning their bodies can deliver more oxygen to muscle cells faster.

The geese's flight muscles also have more mitochondria—energy-producing structures inside cells—than their fellow fowl. Another trick in the birds' arsenal: hyperventilation. Unlike humans, bar-headed geese can breathe in and out very rapidly without getting dizzy or passing out. "By hyperventilating, [the birds] increase the net quantity of oxygen that they get into their blood," Hawkes explained.

> "We will never be able to engineer a human lung to work like a bird lung. But with more information we might be able to develop drugs that would help duplicate some of their cellular responses."
> **Frank Powell**
> *professor of medicine and director of the University of California's White Mountain Research Station*

Older Than the Hills

Before the new study, many scientists had thought the geese were taking advantage of daytime winds that blow up and over mountaintops. But the team showed the birds forgo the winds and choose to fly at night, when conditions are relatively calmer.

"They're potentially avoiding higher winds in the afternoon, which might make flights more uncomfortable or more risky," said Bishop. The birds could potentially head east or west and fly around, rather than over, the mountain range, but this would add several days to their trip and would actually use up more energy.

"I guess it's like when you're trying to get in the grocery store and there's a steep set of stairs and a really long [wheelchair] route. You have to work out which one you can be bothered to do on a particular day," Hawkes said.

Another hypothesis about why the geese choose to fly over rather than around the Himalaya is that the birds have been doing so for millions of years—long before the mountains reached their current heights. "Geese are a relatively old group of birds, and it's possible that when bar-heads first evolved as a species, the mountains weren't nearly as high as they are today," Hawkes said. ∎

Body Odor
Attracting Predators to Birds

New Zealand's native-bird B.O. is so pungent, it's alerting predators to the birds' presence, ongoing research shows.

At best, body odor is offensive, but at its worst, it can be deadly. New Zealand's native bird populations are being threatened, and scientists believe their pungent smell may be exposing them to alien predators. The smells may drive some species to extinction, unless conservationists take unorthodox measures, such as adding "deodorant" to bird nests, according to biologist Jim Briskie of Canterbury University in Christchurch, New Zealand.

Bird Smell Source

Many bird scents stem from a gland that produces waxes essential to keeping feathers healthy. In Europe and the Americas, birds' bodies alter this preening wax during breeding season, changing the wax's composition to reduce smells and keep the birds' nests less detectable by predators that use their noses to find food.

TRUTH:
THERE ARE ONLY ABOUT
62 KAKAPO PARROTS
LEFT IN THE WORLD.

In a recent experiment in New Zealand, Briskie compared waxes from six native species, such as robins and warblers, with waxes from invasive species, such as blackbirds and sparrows, which had evolved in Europe until the 1870s.

"The European birds in New Zealand changed their preen waxes to become less smelly in the breeding season," he said. "But native birds did not, and they remained more smelly overall," Briskie said.

For instance, native kiwis—flightless, chicken-size birds—smell like ammonia, and kakapo parrots, also flightless, smell like "musty violin cases,"

The kakapo parrot's strong body odor may be life threatening.

he said. Other New Zealand species seem to have similarly distinctive scents, Briskie said, unlike most birds on other continents.

"We do know that it's easy for muzzled dogs to find kakapo and kiwi by their smell, so I suspect that predators like rats or feral cats might be able to easily find native birds also," Briskie said.

Alien Predators on the Scent

New Zealand's birds may be so pungent largely because they were able to get away with it for so many centuries, Briskie suspects. When New Zealand split away from Australia some 80 million years ago, no predatory mammals came along for the ride, so native birds never had to evolve means of masking their scents to survive, he said.

But eventually humans changed the landscape. The native Maori people introduced the Polynesian rat, and Europeans later unleashed other rat species, domestic cats, and the stoat—a type of weasel—which have easily caught on to the birds' scents.

Partly as a result, some 43 native birds have already gone extinct, Briskie said.

TRUTH:
KIWI BIRDS ARE THE SIZE OF CHICKENS BUT LAY EGGS THE SIZE OF OSTRICH EGGS.

Seventy-three other native species, many of them flightless, are listed as threatened by the International Union for Conservation of Nature.

Deodorant for Birds

One solution could be to give deodorant to the odorous birds, Briskie said. "If we prove that this is a problem, we might be able to envision some kind of odor-eater or deodorant we could put into the nest to absorb some of those odors and protect them more effectively," Briskie said.

But there's a potential downside—the birds' stench may serve other purposes. Bird deodorant "would only be useful as long as we knew it didn't interfere with the way those odors might be used in communication with mates or off-spring," Briskie explained.

In addition, bird B.O. might also be used to turn the tables on predators. "It could be another way of building a better mousetrap to catch [invasive] rats or stoats," Briskie said. "Perhaps instead of controversial poisons, we might come up with long-lasting baits using essence of kiwi or kakapo that lure predators into a trap." ∎

The Smell of Love

Bird experts have long denied the importance of smell in birds' communication patterns, focusing more on birdsongs and flashy plumage. However, a recent study on the crested auklet shows that smell may be as important to birds as to other animals. The Alaskan seabird produces a strong tangerine-like smell, which scientists associate with courtship displays. The fragrance may act as a type of sexual ornament, since crested auklets already display other traits such as bright beaks that are used to attract mates or show status. The strength of a bird's smell may indicate quality to a mate, although this theory requires further research.

Half-Male, Half-Female Chicken

Call them "hoosters" or call them "rens," but scientists have finally discovered the reason why some chickens are half male and half female.

It was a tough egg to crack, but scientists have finally explained why some chickens are born half male and half female. The bodies of these hen-rooster hybrids, or gynandromorphs, have a mixture of genetically male and female cells, the research reveals.

Appearances Can Be Confusing

Only about 1 in 10,000 chickens are born as gynandromorphs, which have male features—such as a rooster's comb and a defensive leg spur—on one side of their bodies and dainty, henlike features on the other.

Researchers had thought a rare genetic abnormality caused the condition. To test this theory, Michael Clinton of the University of Edinburgh and his team analyzed cells from three gynandromorph chickens. To their surprise, the team found that the chickens' cells were normal. What was strange, however, was that male cells made up one half of the body, and female cells composed the other half.

> **TRUTH:**
> CHICKENS SEE DAYLIGHT 45 MINUTES BEFORE HUMANS DO.

A hybrid chicken—reflected in mirrors—has both male (left) and female (right) characteristics.

Eggs Are Double Fertilized

The scientists believe gynandromorphs are created when a chicken egg becomes fertilized by two sperm. Despite their dual nature, the hybrid birds typically have one of the sex organs, either testes or ovaries. The scientists did not test whether the chickens could actually reproduce, however.

Gynandromorphs are known to exist in other bird species, such as zebra finches, pigeons, and parrots, Clinton said by email. It's likely that the phenomenon occurs in all bird species, he added, but it's not always obvious because males and females of many species often look similar. ■

Sexually Showy Male Birds

Finish Early

Among male houbara bustards, large brown birds found in North Africa, wooing females can be a tricky business because, for them, the more mating dances they do, the lower their sperm counts.

Live fast, age fast—at least if you're a male houbara bustard. That's because male bustards that perform longer courtship displays lose sperm quality faster than males that do not put on elaborate seduction shows, a new study suggests.

The sheer energy required to keep up marathon performances eventually takes its toll on the sperm production of the flashy males, which actually start out with healthier, more robust sperm than their humdrum rivals.

> **TRUTH:**
> HOUBARA BUSTARDS ARE HUNTED FOR THEIR MEAT, WHICH IS THOUGHT TO BE AN APHRODISIAC.

Risky Business

"In nature, life is very risky, so you need to balance the benefit that certain behaviors can give you at present, and the costs that these same behaviors can incur later on," said study senior author Gabriele Sorci, an evolutionary biologist at the University of Burgundy in France.

In the case of the bustards, Sorci thinks the showy males get the balance right, in terms of survival. In part because of the high chances of dying young in the wild, he said, "It's always the best [reproduction] strategy to have early benefits and eventually pay a later cost." The new study represents the first time that such a trade-off has been linked to declines in male fertility in any species, Sorci added.

The Dance of Love

In their North African habitat, male houbara bustards perform long displays to attract females up to six months out of the year. Some males keep at it for several hours a day and perform more often, while others invest less of their resources to luring mates.

After strutting for a while, a male erects an ornamental "shield" of long white feathers and then runs at high speed, often circling a rock or a bush, according to the study, published recently in the journal *Ecology Letters.* The show climaxes in a flash of black and white feathers and several booming calls so deep they're almost out of range of human hearing, according to Sorci. Females will often select males that run more laps while taking fewer and shorter breaks.

Shooting Blanks

For the study, the team used ten years of data taken from more than 1,700 male bustards—ranging from 1 to 24 years old—at captive-breeding facilities in Morocco. Each day, workers would observe the males' courtship behavior,

A male houbara bustard performs his mating dance.

which is roughly the same as in the wild. The scientists then added up the number of days the males were seen displaying, and for how long. The result was an index of "male sexual-display effort" for every year of each bird's life.

Also daily, a team would place a dummy female under each male bird to initiate mating and then capture his ejaculate in a petri dish, the study says. Scientists recorded the quality of the bird's semen—how many sperm were in each ejaculate, how well they swam, etc.

The results showed that the most avid performers during youth released smaller quantities of semen, with more dead and abnormal sperm, at older ages. The data also uncovered a still unresolved mystery: Though these flashy males had passed their reproductive prime, the show still went on, Sorci noted.

Fabrice Helfenstein, an evolutionary ecologist at the University of Bern in Switzerland, said via email that the study "is sound—it is based on a lot of animals and uses properly new statistical tools." For instance, the captive birds proved useful: "Such aging processes are usually hard to reveal in wild animals," which are generally thought to die too young to suffer the effects of aging, said Helfenstein, who was not involved in the new study.

Like Human, Like Bird

Male great tit birds with brighter breasts also seem to have stronger sperm, a recent study finds. This advertisement is a boon to female great tits, since finding a male with high-quality sperm isn't always easy. This is partially due to free radicals, which threaten sperm cells in many animals, including humans. Male great tits produce an antioxidant called carotenoid that both defends against free radicals and gives their breast feathers a bright yellow hue. Similarly, some studies have shown that men with more attractive faces have better quality semen.

Casanovas, Take Note

In general, the idea that investing in sexually attractive traits early in life racks up costs later could possibly be applied to other species, including humans, study co-author Sorci said. For instance, showy male bustards may be the "bird equivalent of the posers who strut their stuff in bars and nightclubs every weekend," study leader Brian Preston, also of the University of Burgundy, said in a statement. "If the bustard is anything to go by, these same guys will be reaching for their toupees sooner than they'd like." ■

Why Do Birds Fall From the Sky?

In-air deaths of large groups of birds have spooked people for centuries, but what is behind the most recent rash of these seemingly ominous events?

A mysterious rain of thousands of dead birds darkened New Year's Eve 2010 in Arkansas, and in early January 2011 similar reports streamed in from Louisiana, Sweden, and elsewhere. But the in-air bird deaths aren't due to some apocalyptic plague or insidious experiment—they happen all the time, scientists say. The 2011 buzz, it seems, was mainly hatched by media hype.

At any given time there are "at least ten billion birds in North America . . . and there could be as much as 20 billion—and almost half die each year due to natural causes," said ornithologist Greg Butcher, director of bird conservation for the National Audubon Society in Washington, D.C.

But what causes dead birds to fall from the sky en masse? The Arkansas case points to two common culprits: loud noises and crashes.

> **TRUTH:**
> **DRAMATIC DIE-OFFS ARE ACTUALLY VERY COMMON IN ANIMALS THAT CONGREGATE OR TRAVEL IN LARGE GROUPS.**

New Year's Eve, 2010

Beginning at roughly 11:30 p.m. on New Year's Eve, Arkansas wildlife officers started hearing reports of birds falling from the sky in a square-mile area of the city of Beebe. Officials estimate that up to 5,000 red-winged blackbirds, European starlings, common grackles, and brown-headed cowbirds fell before midnight.

Results from preliminary testing released by the National Wildlife Health Center in Madison, Wisconsin, show the birds died from blunt-force trauma, supporting preliminary findings released by the Arkansas Livestock and Poultry Commission.

"They collided with cars, trees, buildings, and other stationary objects," said ornithologist Karen Rowe of the Arkansas Game and Fish Commission. "Right before they began to fall, it appears that really loud booms from professional-grade fireworks—10 to 12 of them, a few seconds apart—were reported in the general vicinity of a roost of the birds, flushing them out," Rowe said.

> # TRUTH:
> ## FIVE BILLION BIRDS DIE IN THE UNITED STATES EVERY YEAR.

"There were other, legal fireworks set off at the same time that might have then forced the birds to fly lower than they normally do, below treetop level, and [these] birds have very poor night vision and do not typically fly at night."

The dead birds found in Arkansas are of a species that normally congregate in large groups in fall or winter. "The record I've heard is 23 million birds in one roost," Audubon's Butcher said.

"In that context, 5,000 birds dying is a fairly small amount."

A Towering Problem for Birds

Birds often hit objects in flight, especially "tall buildings in cities, or cell phone towers, or wind turbines, or power lines," Butcher said. "The structures that seem to cause the most deaths are very tall and constantly lit," he said. "On foggy nights, birds that should probably normally be paying attention to the stars get disoriented, and circle around the structures until they collapse" and fall.

Collisions with power lines seem to have killed roughly 500 blackbirds and starlings in Louisiana in early January 2011. The 50 to 100 jackdaws found on a street in Sweden that same day showed no signs of disease and also apparently died from blunt-force trauma, according to the Swedish National Veterinary Institute.

Wind, snow, hail, lightning, and other challenges posed by weather can easily kill flying birds too. For example, "last year a couple of hundred pelicans washed up by the Oregon-Washington border," Butcher said. "A cold front had unexpectedly moved in, and they faced icing on their wings and bodies."

What Are the True Crises?

Of course, death doesn't just stalk birds from above. For instance, "waterfowl get botulism—and salmonella and avian pox can spread at bird feeders," Butcher said.

No matter how it arrives, death appears to be very much a fact of life for birds. "Young birds that hatch in the spring have an approximately 75 percent chance of not reaching their first birthdays," the Arkansas Game and Fish Commission's Rowe said.

To biologists, these deaths are normal occurrences. "I wish I could take all this energy and attention on these deaths and direct them toward true crises in wildlife biology, to things like the white-nose syndrome in bats," Rowe added.

She does, though, see a silver lining in the sky-is-falling coverage in 2011. "I hope we can raise public awareness of what impact man-made structures can have on other species. How many migratory warblers do you want to kill just to get better cell phone reception?" ∎

> **"In Arkansas . . . fireworks were set off in a town near a known blackbird roost. [The] birds flushed from the roost and . . . were seen crashing into buildings and cars and poles. Necropsies show blunt force trauma to brain and breast."**
>
> **Melanie Driscoll**
> *biologist and director of bird conservation for the Gulf of Mexico and Mississippi Flyway for the National Audubon Society*

Bird With "Human" Eyes

Knows What You're Looking At

The messages in our looks, glances, and gazes might be no mystery at all to the jackdaw, a blackbird with humanlike eyes.

For the crowlike birds known as jackdaws, it's all in the eyes. The species may be the only animal aside from humans known to understand the role of eyes in seeing and perceiving things, according to a new study.

The Eyes Have It

While humans often use visual clues to communicate, it wasn't known whether other animals share this social ability. Jackdaw eyes, like those of humans, are unusually conspicuous, with dark pupils surrounded by silvery white irises.

The physical similarities hint that jackdaws use their eyes to communicate in the same ways humans do, said study leader Auguste von Bayern, a zoologist currently with the University of Oxford. "We can communicate a lot via the eyes, and jackdaws do that as well, in my opinion," von Bayern said.

Now her study of hand-reared jackdaws shows that the birds—members of the same family as crows and ravens—can use a human's gaze to tell what that person is looking at. "They

> **TRUTH:**
> **A BIRD'S VISION IS ITS MOST IMPORTANT SENSE.**

A Eurasian jackdaw gives someone the eye.

are sensitive to human eyes because they are sensitive to their own species' eyes," von Bayern said. By contrast, previous studies have shown that other animals regarded as intelligent, such as chimpanzees and dogs, find even their own species' eyes hard to read.

Conflict and Cooperation

Von Bayern conducted the jackdaw experiments while completing her Ph.D. at the University of Cambridge. In one test, she and colleague Nathan Emery timed how long a jackdaw took to retrieve food if a person was also eyeing the prize.

They found that the birds took longer to retrieve the food if the human was unfamiliar—someone the bird apparently didn't trust. The birds were equally sensitive to the gaze of a single eye, such as when the person looked at the food in profile or kept one eye closed. This suggests the jackdaws made the decision to risk conflict solely based on eye motion and not on other cues, such as the direction a potential rival's head was facing.

In a second experiment, the birds were able to interpret a familiar human's altered eye gaze to "cooperate" to find food that was hidden from view. The study authors add that more tests will be needed to tell if the birds were able to read eye movements based on their natural tendencies or if it is a learned behavior from being raised by humans. ■

Bird "Sings"
Through Feathers

Most birds twitter and chirp through their beaks, but the feathers of this South American songbird make a music all their own.

Solving a long-standing puzzle among bird experts, scientists have found that the sharp, violin-like sounds of a South American songbird come not from the beak but from a suite of specially evolved, vibrating feathers.

> "I was just utterly stunned. There's literally no bird in the world that does anything that prepares you for it. It's totally unique."
>
> **Richard Prum**
> *ornithologist, Yale University, on hearing a club-winged manakin "sing" with its feathers for the first time*

Club-Winged Manakin Music

A new study offers the first hard evidence that birds use feathers for audible communication as well as for flight and warmth. In 2005 Kimberly Bostwick theorized that the male club-winged manakin—a tiny bird of the Andean cloud forest—was vibrating a club-shaped wing feather against a neighboring, ridged feather to "sing" when trying to attract females. Proving the feather-song connection, though, would be a huge challenge.

"It was very hard to mess with the birds' feathers and still have them do their display," said Bostwick, curator of birds and mammals at the Cornell University of Vertebrates in Ithaca, New York. "There were many times where

I listened to the sound and started doubting that a feather could possibly make [the sound]," she recalled.

Bird Vibrations

To determine, once and for all, how the manakin was making its bizarre sounds, Bostwick and colleagues decided to take feather samples and analyze them in a lab.

She knew from previous work that the frequency of the sound made by the manakin was 1500 hertz —1,500 cycles per second. If the two feather types were making the sound, they should resonate when vibrated at the same frequency during the experiments. The team used lasers to monitor vibrations as they were oscillated by a lab device called a mini-shaker. The special feathers vibrated at exactly 1,500 hertz—proving they're responsible for the strange sounds.

But there's a twist: Bostwick was surprised to find that the club and ridged feathers aren't a duet, but part of a chamber orchestra. Individually the manakin's "regular" feathers didn't resonate like the special ones. But when the nine feathers closest to the special feathers were still attached to the ligaments, they vibrated at around 1500 hertz, harmonized with the club feathers, and amplified the volume of the sound.

The results, Bostwick said, could lead to better understanding of the newly discovered form of bird communication. "Lots of birds make simple clapping sounds or whooshing noises with their wings, and we haven't even begun to understand how the sounds are made or how they've evolved," she added. ■

> **TRUTH:**
> THE WAY A CLUB-WINGED MANAKIN RUBS ITS FEATHERS BACK AND FORTH TO CREATE SOUND IS SIMILAR TO HOW CRICKETS MAKE CHIRPING NOISES.

Birds Can "See" Earth's Magnetic Field

To find north, humans look to a compass. But birds may just need to open their eyes, a new study says.

A bird's view of the world may include more than just sweeping vistas and grand landscapes. A new study indicates that it might just include Earth's magnetic field.

Scientists already suspected birds' eyes contain molecules that are thought to sense Earth's magnetic field, but not to actually see it.

In a new study published in the Public Library of Science journal *PLoS ONE*, German researchers found that these molecules are linked to an area of the brain known to process visual information. In that sense, "birds may see the magnetic field," said study lead author Dominik Heyers, a biologist at the University of Oldenburg.

TRUTH: BIRDS CAN SEE TWO TO THREE TIMES BETTER THAN HUMANS.

Magnetic Orientation

Man-made compasses work by using Earth as an enormous magnet and orienting a tiny magnet attached to a needle to the planet's north and south poles. Scientists have thought for years that migratory birds may use an internal compass to navigate between their nesting areas and wintering grounds, which can be separated by thousands of miles. The new research helps explain how this natural compass may work.

Heyers and his colleagues injected migratory garden warblers with a special dye that can be traced as it travels along nerve fibers. The team put one type of tracer dye into the eyes and another in a region of the brain called Cluster N, which is most active when birds orient themselves. When the birds got their bearings, both tracers traveled to and met in the thalamus, a region in the middle of the brain responsible for vision.

"That shows there is direct linkage between the eye and Cluster N," Heyers said. The finding strongly supports the hypothesis that migratory birds use their visual system to navigate using the magnetic field. "The magnetic field or magnetic direction may be perceived as a dark or light spot which lies upon the normal visual field of the bird," Heyers said, "and which, of course, changes when the bird turns its head."

More Navigational Tools

Scientists not involved with the study said it is impressive and well done, but cautioned that there are more pieces to the puzzle of how birds navigate on their long migrations.

"An animal that has to migrate over great distances needs to have both a compass and a map," said Cordula Mora, a biologist who recently completed her postdoctoral research at the University of North Carolina, Chapel Hill. Mora's work suggests that birds may use magnetic crystals in their beaks to sense the intensity of the magnetic field and thus glean information on their physical location.

"If you have a compass, you know where north, south, east, [and] west [are], but you don't know where you are, so you don't know where you should be going," she said.

Damaged Magnetic Field Cause of Bird Die-Offs?

If birds cannot see the magnetic field when migrating, they lose their bearings and could hurt themselves or die. In 2008, NASA reported a "massive breach" of the Earth's magnetic field. Solar wind can flow into the breach, producing large geomagnetic storms. In addition to this being a possible reason for why thousands of birds across the planet are dying suddenly, it may also cause lost radio transmissions, radiation from too much solar power, and changes in the Earth's crust, which could lead to increases in landslides, mudslides, earthquakes, and volcanic eruptions.

A migratory garden warbler

Study author Heyers said "both [map and compass] systems may act in concert."

Robert Beason is a wildlife research biologist with the U.S. Department of Agriculture in Sandusky, Ohio, and an expert on bird navigation. He noted that stars may also either fully or in part provide the birds with their visual bearing—not the magnetic field.

The next step is to figure out where all this information comes together in the bird brain, he noted. "That's probably going to tell us where the navigation center for birds is," he said. ■

Chubby Snipe

Snaps Nonstop Record

A plump little shorebird has smashed the record for the fastest long-distance, nonstop flight in the animal kingdom, making a trip from Sweden to sub-Saharan Africa in two days.

Don't let the chubby snipe's name fool you. This is one fast moving bird. In a new study, scientists have discovered that great snipes can complete a transcontinental flight across Europe, from Sweden to sub-Saharan Africa, in as little as two days without resting. The birds traveled up to 4,200 miles (6,760 kilometers) at an average speed of 60 miles (97 kilometers) an hour.

Fast Fat Fliers

To track the birds, biologists captured and tagged ten great snipes with geolocators at their breeding grounds in western Sweden in May 2009. Tracking data for three of the birds was retrieved after their recapture in Sweden a year later.

At first glance, great snipes don't look especially speedy or well equipped for such an arduous journey. Their bodies are small and chubby, not aerodynamic, and in the autumn the birds get

> "If we understand what sleep is doing to their brains then we may be able to manipulate the neurochemistry in the human brain to do something similar someday."
>
> **Jerome Siegel**
>
> *neuroscientist and professor of psychiatry, UCLA, on understanding how birds operate on so little sleep during migrations*

so rotund that, when shot, "their skin sometimes ruptures when the shot bird hits the ground," according to one 19th-century account.

But it's these ample fat reserves that allow the birds to fly such long distances without stopping, said study leader Raymond Klaassen, a biologist at Sweden's Lund University. "They almost double their body weight before the flight," Klaassen said. "And all this fat will be burned during the flight, and they will arrive lean and exhausted in Africa."

All-Nighters

Birds are known to handle sleep deprivation better than any other mammal. For example, pigeons can survive for months with only about 10 percent of their normal sleep time. One hypothesis for migrating birds' ability to sleep so little was unihemispheric sleep, where half the brain sleeps and the other half remains active. Dolphins and fur seals are unihemispheric sleepers—they swim and breathe while literally half asleep. However, after preliminary experiments on Swainson's thrushes, scientists found that unihemispheric sleep was probably not the primary way the birds handled sleep loss. The most intriguing observation was that the thrushes were less active and took more naps during migration season.

Speed and Endurance

It's a rare bird that can fly both far and fast. For example, the Arctic tern racks up as many as 50,000 miles (80,000 kilometers) during its yearly migration from the Arctic to the Antarctic and back again. However, the animal spreads the flight out over several months and fishes along the way. At the other end of the spectrum, peregrine falcons can reach speeds of up to 200 miles (322 kilometers) an hour but only in short bursts to catch prey.

The only other bird that comes close to matching the great snipe's abilities is the godwit, a wading bird. In 2007 scientists recorded a godwit flying more than 7,000 miles (11,500 kilometers), from Alaska to New Zealand, in nine days at an average speed of about 35 miles (56 kilometers) an hour. "[One] difference between the godwits and the snipes is that the godwits travel over the ocean, and thus have no possibilities to stop," Klaassen said. "Hence, their amazing flights are not their choice."

By contrast, snipes have several rest-stop options during their autumn migration to Africa but choose not to take advantage of them. The reasons for that are unclear, especially since the birds make several stops during their return flights to Sweden in the spring.

Bird Migration "Revolution"

As with other migratory birds that fly long distances, it's unclear how great snipes can apparently fly for such long periods with little or no sleep. "This is one of the unsolved mysteries of long-distance flights," Klaassen said. "We now believe that half of their brain sleeps at [a] time, alternating between the left and the right side. Or they do not sleep at all, but this seems impossible regarding the importance of sleep in general."

Klaassen says so little is known about bird-migration strategies that he wouldn't be surprised if the great snipe's record is broken soon. "Generally we know rather little about the performances of different species, as many have not yet been tracked," he said.

"I foresee many surprises in the nearby future, due to the recent development of minute tracking devices. The field of bird migration is currently going through a revolution, and these are certainly very exciting times for us." ∎

TRUTH: GREAT SNIPES' SPEED DOES NOT RELY ON WIND ASSISTANCE, WHICH MAKES IT EVEN MORE IMPRESSIVE.

Elusive "Smiling" Bird

Captured on Film

The notoriously camera shy recurve-billed bushbird put on a happy face for his first photograph in 40 years.

Call him the Mona Lisa of the bird kingdom. The rare recurve-billed bushbird, recently rediscovered by scientists in Colombia after a 40-year absence, sports a curving beak that gives the illusion of an enigmatic smile.

Say cheese! The rare recurve-billed bushbird poses for the camera.

Ready for His Close-Up

The photograph, taken by a conservationist with the Colombia-based nonprofit Fundación ProAves in 2005, is the first photo released of a live bushbird. The elusive species had not been spotted between 1965 and 2004, due to its limited range and remote habitats. It was seen recently in Venezuela and in a region of northeastern Colombia, where it was photographed.

Researchers found the bird in a 250-acre (101-hectare) reserve next to the Torcoroma Holy Sanctuary near the Colombian town of Ocaña, where in 1709 locals claimed they saw the image of the Virgin Mary in a tree root. The forests of the sanctuary have been protected by Catholic Church authorities in the centuries since.

Another Rare Bird

The researchers also found and photographed the extremely rare Perija parakeet, of which only 30 to 50 individuals likely survive. Deforestation and wildfires for agriculture and grazing have denuded much of the birds' habitat, conservationists say.

"[A]s more and more remote areas are being settled, the bushbird reminds us how important it is to conserve as much natural habitat as we can," said Paul Salaman of the American Bird Conservancy. "Who knows what wonderful biodiversity is being destroyed before it has had a chance to be discovered?" ■

> **TRUTH:**
> THE BUSHBIRD USES ITS UNIQUE BILL TO SLICE OPEN THIN BAMBOO SHOOTS AND DIG OUT BEETLE LARVAE AND OTHER ARTHROPODS FROM THE STALK.

Superfast Muscle Power

Found in Songbirds' Throats

Faster than a speeding . . . songbird? The world's fastest moving muscles don't belong to Olympic sprinters, they're in the throats of songbirds.

The fastest muscles known lie within the throats of songbirds, according to new research on how birds vibrate their vocal cords. The study, published in the journal *PLoS One*, shows that the muscles used by birds to create their songs are the speediest of all.

Strong Songs

Experts have long wondered whether bird song is caused by passive interactions as air moves between the vocal muscles or direct neuromuscular control.

"I had been looking at the muscles in a pigeon species and was amazed by how fast they were moving," said lead study author Coen Elemans at the University of Utah in Salt Lake City. "[Pigeons] have really boring, slow songs, and it made me wonder what the muscles in songbirds were like, so I decided to find out."

What Elemans and colleagues discovered is that zebra finches and European starlings can change their tunes at frequencies as high as 250 hertz via direct muscle control. This means that they are moving their muscles a hundred times faster than a blink of the human eye.

Extremely Fast Twitch

To find out how songbirds make their quick-fire modulations, the researchers first measured muscle activity in freely singing starlings and found that muscle motion corresponded to changes in song tone.

The team then exposed vocal muscle fibers from starlings and zebra finches to electrical stimulation in the lab to see just how fast the muscles can expand and contract. The vocal muscles of male and female starlings both contracted at about 3.2 milliseconds. Male zebra finch muscles, meanwhile, twitched at roughly 3.7 milliseconds while females' moved at 7.1.

Tiny twitching muscles on either side of rattlesnake rattles, along with muscles in the swim bladders of some fish, have been recorded approaching these speeds. But Elemans's team concludes that songbird vocal cords move faster than any muscle in any other known vertebrate.

Since most songbirds have the same general type of vocal cords, the discovery could mean that extremely fast-moving muscles are more common in nature than was previously thought.

> "Superfast muscles were previously known only from the sound-producing organs of rattlesnakes, several fish and the ring-dove. We now have shown that songbirds also evolved this extreme performance muscle type, suggesting these muscles— once thought extraordinary— are more common than previously believed."
>
> **Coen Elemans**
> *biology researcher,*
> *University of Southern Denmark*

Synchronized Chorus

Daniel Margoliash, a biologist at the University of Chicago who was not involved with the study, called the paper "as elegant as it is exciting." "We've been fascinated by bird songs for so long, and this gives us a very important

TRUTH:
MALE NORTHERN MOCKINGBIRDS MAY LEARN AS MANY AS 200 SONGS THROUGHOUT THEIR LIVES.

insight into the vocal organs behind them," he said. "We had no idea muscles could work at these superfast rates," he added. "That they can and do is just amazing."

Daniel Mennill, an avian biologist at the University of Windsor in Canada, noted that fieldwork has shown songbird vocalizations to be among the most precisely timed behaviors in the animal kingdom. "The synchronized duets and choruses of wrens, for example, are the most highly coordinated animal behaviors ever recorded," he said. "These [new] results explain a lot about how birds actually achieve such amazing technical feats."

Study leader Elemans said he is keen to continue his search for creatures with superfast muscles, and he thinks bats will be good candidates. "Bats echolocate with an auditory sweep that rapidly moves from very high pitch to very low," he said. "I'm convinced that there are fast-moving muscles behind this sonic sweep." ■

Crows Have Human-Like Intelligence
Author Says

Clever crows are one of nature's brainiest birds, sharing the similar hallmarks of higher intelligence with humans.

Crows make tools, play tricks on each other, and caw among kin in a dialect all their own. These are just some of the signs that point to an unexpected similarity between the wise birds and humans. "It's the same kind of consonance we find between bats that can fly and birds that can fly and insects that can fly," said Candace Savage, a nature writer based in Saskatoon, Canada.

"Species don't have to be related for there to have been some purpose, some reason, some evolutionary advantage for acquiring shared characteristics," she added.

Brainy Black Birds

Crows are one of the few species that use tools.

Author of *Crows: Encounters with the Wise Guys of the Avian World,* Savage explores the burgeoning field of crow research, which suggests that the birds share with humans several hallmarks of higher intelligence, including tool use and sophisticated social behavior.

Sharing Intel

In a five-year study of crows living near Seattle, researchers discovered that the crows can remember a "dangerous human" and are able to share their knowledge of the danger with their offspring and other crows. "Because human actions often threaten animals, learning socially about individual people's habits would be advantageous," says study co-author John Marzluff of the University of Washington. "Crows recruit and tolerate others of their own and different species in mobs that form around dangerous people. This social tolerance could allow naïve crows to learn about dangerous situations, locations and individual humans."

The shared traits exist despite the fact that crows and humans sit on distinct branches of the genetic tree. Humans are mammals and crows are birds, which Savage calls feathered lizards, referring to the theory that birds evolved from dinosaurs.

"I'm not positing there's anything mythological about this or imagining crows are in any way human," she said. "But whatever it is that has encouraged humans to develop higher intelligence also seems to have been at work on crows."

Tool Use

Alex Kacelnik is a zoologist at Oxford University in England who studies tool use in crows. He said study of the birds advances understanding of how higher intelligence evolves.

As a sign of crows' advanced smarts, Savage cites Kacelnik's 2002 study in the journal *Science* on a captive New Caledonian crow that bent a straight piece of wire into a hook to fetch a bucket of food in a tube. "No other animal—not even a chimp—has ever spontaneously solved a problem like this, a fact that puts crows in a class with us as toolmakers," Savage writes in her book.

Kacelnik noted that New Caledonian crows, which are restricted to a few islands in the South Pacific Ocean, are the only example of some 45 crow species that "are very intense tool users in nature.'" Nevertheless, he continued, these birds are "both intense tool users and creative tool users . . . In addition to the tools they are normally seen to use in the wild, they are capable of making new instruments when the necessity arises," such as the wire hooks.

In research published in the journal *Nature,* Kacelnik and his colleagues demonstrated that New Caledonians are born toolmakers—that there is a genetic component to the behavior. The finding, Kacelnik said, fits the notion that higher intelligence requires a genetic imprint to foster more advanced behaviors like learning and innovation.

"There are three elements: what animals inherit, what animals learn by individual experience, and what animals acquire through social input," he said. "It's a mistake to believe [these elements] compete. Actually, they coalesce, they enhance each other."

Crow Trickery

The intelligence of other crow species, most notably ravens, is also demonstrated by their ability to manipulate the outcomes of their social interactions, according to book author Savage.

> **TRUTH:**
> CROWS ARE VERY SOCIAL AND LIVE IN FAMILY GROUPS CONSISTING OF BETWEEN 2 AND 15 BIRDS, WITH 4 BEING THE AVERAGE.

For example, she highlights raven research by University of Vermont zoologist Bernd Heinrich showing how juvenile and adult ravens differ when feeding on a carcass. The juveniles cause a ruckus when feeding to recruit other young ravens to the scene for added safety against competition with adult crows and other scavengers. The adults, by contrast, show up at a carcass in pairs and keep quiet to avoid drawing attention—and competition—to the food.

Savage also discusses Swiss zoologist Thomas Bugnyar's research showing how a raven named Hugin learned to deceive a more dominant raven named Mugin into looking for cheese morsels in empty containers while Hugin snuck away to raid full containers.

"This shady behavior satisfies the definition of 'tactical,' or intentional, deception and admits the raven to an exclusive club of sociable liars that in the past has included only humans and our close primate relatives," Savage writes in her book.

Another area of crow research that may indicate higher intelligence is how crows learn and use sound. Preliminary findings suggest that family groups develop their own sort of personal dialects, according to Savage. "There's a lot more going on in a bird brain than people ten years ago would have imagined," she said. ■

Why Transylvanian Chickens Have Naked Necks

Some say "churkeys," some say "turkens," but science says that a genetic mutation is behind the bare-naked necks of Transylvanian chickens.

Scientists have cracked why the Transylvanian naked neck chicken has a featherless neck—and it isn't to give vampires easier access. The Transylvanian bird's bare neck results from a random genetic mutation that causes the overproduction of a feather-blocking molecule called BMP12, a new DNA study shows.

> **TRUTH:**
> MIKE THE CHICKEN LIVED FOR 18 MONTHS WITHOUT A HEAD, FROM 1945 TO 1947.

Churkeys and Turkens

The mutation first arose in domestic chickens in northern Romania hundreds of years ago. The naked neck chicken—also dubbed the churkey or turken—has a chickenlike body but a turkeylike head atop a long, deep-red neck.

Surprisingly, when scientists treated standard-breed chicken embryos with BMP12 in the lab, the young chickens developed no feathers on their necks—suggesting the neck is more sensitive to the molecule, according to study leader Denis Headon, a developmental biologist at the University of Edinburgh's Roslin Institute.

Transylvanian naked neck chickens' vampire-friendly look is genetic.

To find out why, Headon and his team did a further analysis, which revealed that an acid derived from Vitamin A is produced on the chicken's neck skin. The acid essentially enhances the BMP molecule's effects, making the birds' necks bare, they found.

Naked Necks, Cooler Birds

Unlike most genetic mutations, which are generally bad for an animal, the naked-neck tweak has increased naked necks' popularity worldwide. That's because bare-necked birds are more resistant to heat and thus produce better meat and eggs—especially crucial for poultry producers in hot climates such as Mexico, Headon said.

And naked necks aren't alone: "We think all birds have this priming or readiness to lose neck feathers first," he noted. "Once you have a mutation that increases BMP12 in skin, the neck is the region that's ready to lose its feathers—it's already more sensitive."

In the wild, for instance, it's likely that ostriches and storks have lost their neck feathers to stay cool, though it's unclear whether BMP12 played any role. "Evolution has always found it easy to lose neck feathers whenever it gets hot and the bird gets big." ∎

CHAPTER 7

The Final Fron

tier

A long time ago, in a galaxy far, far away . . . there were some weird things happening. They are still happening—in our own galaxy and in many, many others. There are the cannibal stars that snack on their neighbors, the discovery of a "diamond" exoplanet, and light, fluffy snow found on the moons of Saturn. These things may sound like they're straight out of science fiction, but they're not. They're the cold, hard, weird truth.

What Created Earth's Oceans?

Comet Offers New Clue

Was Earth's water delivered from above? Astronomers believe they know the answer.

For the first time, astronomers have found water on a comet that's a chemical match for water on Earth, a new study says. The discovery backs up theories that water-rich comets helped fill ancient Earth's oceans.

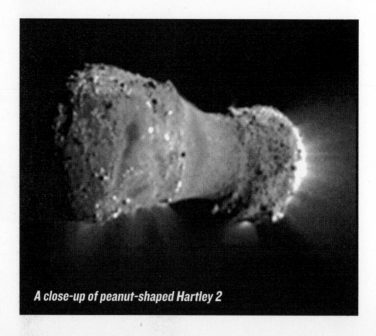

A close-up of peanut-shaped Hartley 2

They Ain't Semiheavy

Planet-formation models indicate that early Earth was much too hot to sustain liquid water on its surface, making the origin of Earth's oceans a mystery. So scientists speculated that our planet's surface water came from comets that slammed into Earth once the planet had cooled.

This theory was dealt a serious blow in the 1980s, however, due to measurements of the ratio of normal to "semiheavy" molecules—the D/H ratio—in comet water. In a semi-heavy water molecule, one hydrogen atom (H) is replaced with a heftier version called deuterium (D). All water in nature has a D/H ratio, and since deuterium is a very stable atom, this ratio can go unchanged for eons.

Since the 1980s, researchers have found that several comets in our solar system have D/H ratios that are very different from that of Earth's water. Those results indicated that, at best, only about 10 percent of Earth's water could have come from comets, with the rest probably coming from water-rich asteroids, explained study leader Paul Hartogh, an astronomer at the Max Planck Institute for Solar System Research in Germany.

Molecular Match

In the new study, Hartogh and his team used the European Space Agency's Herschel Space Observatory to examine the D/H ratio of the comet 103P/Hartley 2. Hartley 2 is a so-called Jupiter family comet because its orbit takes the comet close to the orbits of Jupiter and the other gas giants.

The results show that Hartley 2's water is very similar to that of Earth. Importantly, computer simulations suggest that Hartley 2 originated from the Kuiper belt, a region beyond the orbit of Neptune that is filled with

> "The reservoir of Earth ocean-like material is much larger than we thought, and it encompasses cometary material, which we hadn't recognized. We have to think really hard and try to get a better understanding of what is going in our Solar System, and whether you can really rule out comets as the source of Earth's water."
>
> **Ted Bergin**
> *University of Michigan*

comets and other icy remnants from the formation of our solar system.

This would suggest that the larger group of comets that helped form Earth's oceans originally came from the Kuiper belt. By contrast, the comets with D/H ratios that didn't match Earth's are thought to have originated in the Oort cloud, a reservoir of billions of comets that astronomers think exists far beyond the Kuiper belt.

Other Space Sources

Finding a comet that's a match for Earth suggests more such matches may be out there, which means much more of our planet's water could have come from comets after all, Hartogh said. But exactly how much is still unclear and will require further studies.

"We cannot give a number," Hartogh said. "In principle, all of Earth's water may [have] come from comets. However, it is still possible that a large—or the largest—fraction came from asteroids." ∎

> **TRUTH:**
> IT WOULD TAKE A STACK OF MORE THAN NINE EMPIRE STATE BUILDINGS TO EQUAL THE AVERAGE DEPTH OF THE OCEAN.

New Planet May Be Among Most Earthlike

Weather Permitting

A new planet found about 36 light-years away could be one of the most Earthlike worlds yet—if it has enough clouds, a new study says.

The search for Earthlike planets is one of the most exciting challenges facing astronomers today, and the clouds on one of the latest discoveries, the unpoetically named HD85512b, have captured their attention.

An artist's impression of a "Super Earth" orbiting the sunlike star HD85512b

Liquid Water Likely

HD85512b was discovered orbiting an orange dwarf star in the constellation Vela. Astronomers found the planet using the European Southern Observatory's High Accuracy Radial velocity Planet Searcher, or HARPS, instrument in Chile.

Radial velocity is a planet-hunting technique that looks for wobbles in a star's light, which can indicate the gravitational tugs of orbiting worlds. The HARPS data show that the planet is 3.6 times the mass of Earth, and the new world orbits its parent star at just the right distance for water to be liquid on the planet's surface—a trait scientists believe is crucial for life as we know it.

"The distance is exactly the limit where you want to be to have liquid water," said study leader Lisa Kaltenegger of the Harvard-Smithsonian Center for Astrophysics and the Max Planck Institute for Astronomy. "If you scale it to our system, it's a bit farther out than Venus is to our sun." At that distance, the planet likely receives a bit more solar energy from its star than Earth does from the sun.

Earths Galore?

HD85512b is one of 16 super-Earths that the European Southern Observatory recently discovered using the High Accuracy Radial velocity Planet Searcher (HARPS). Super-Earths are planets whose mass is between one and ten times that of Earth, but whether life exists there requires exploration. Michel Mayor, who led the HARPS team, says, "In the coming 10 to 20 years we should have the first list of potentially habitable planets in the sun's neighborhood. Making such a list is essential before future experiments can search for possible spectroscopic signatures of life in the exoplanet atmospheres."

Cloud Cover

But Kaltenegger and colleagues calculate that a cloud cover of at least 50 percent would reflect enough of the energy back into space to prevent overheating. On average, Earth has 60 percent cloud cover, so partly cloudy skies on HD85512b are "not out of the question," she said.

Of course, clouds of water vapor depend on the presence of an atmosphere similar to Earth's, something that can't be detected on such distant planets with current instruments. Models of planet formation predict that planets with more than ten times Earth's mass should have atmospheres dominated by hydrogen and helium, Kaltenegger said. Less massive worlds—including HD85512b—are more likely to have Earthlike atmospheres, made mostly of nitrogen and oxygen.

Location, Location, Location

So far, the newly detected planet is the second rocky world outside our solar system to be confirmed in its star's habitable zone—the region around a star that's not too hot and not too cold for liquid water. The first, planet Gliese 581d, was previously discovered using the HARPS instrument. This world lies just on the cool edge of its star's habitable zone.

Another promising planet, Gliese 581g, was discovered in 2010 and dubbed the most Earthlike planet yet. But controversy surrounds the claim, with some experts declaring that the entire planet is actually a data glitch. Manfred Cuntz, director of the astronomy program at the University of Texas, Arlington, noted that more information is needed before anyone can speculate whether aliens are wandering around the newfound planet. "It's not their fault no extra information [about the planet's atmosphere] is available right now," Cuntz said of the research team. "It looks like this is a strong candidate, in principle."

> **TRUTH:**
> DAYS ARE LONGER THAN YEARS ON THE PLANET MERCURY.

Potential for Life

In addition to size and location, HD85512b has two other points in its favor for potentially harboring life, Cuntz said. The planet's orbit is nearly circular, which would provide a stable climate, and its parent star, HD85512, is older—and therefore less active—than our sun, which would lower the likelihood of electromagnetic storms damaging the planet's atmosphere.

Not only that, but in principle, the age of the system—5.6 billion years—"gives life a chance to originate and develop," he said. By contrast, our own solar system is thought to be about 4.6 billion years old.

A Great Place for Yoga?

Given current limits on space travel, it's unlikely for now that humans will get to visit HD85512b. But if we could get there, the newfound planet might seem like a fairly alien world: muggy, hot, and with a gravity 1.4 times that of Earth's, study leader Kaltenegger said. On the bright side, "Hot yoga might be one of the things you don't have to pay for there," she quipped. ■

Sun Headed into Hibernation

Solar Studies Predict

There's been a recent flurry of sunspot activity, but scientists believe it's the storm before the calm. The sun may be going on a break.

Enjoy our stormy sun while it lasts. When our star drops out of its latest sunspot activity cycle, the sun is most likely going into hibernation, scientists announced.

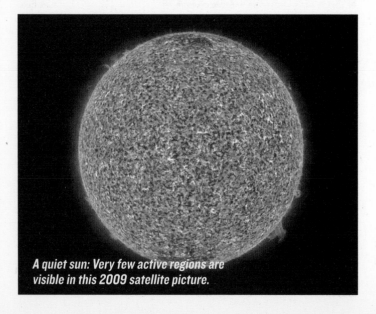

A quiet sun: Very few active regions are visible in this 2009 satellite picture.

Solar Cycle Slow Down

Three independent studies of the sun's insides, surface, and upper atmosphere all predict that the next solar cycle will be significantly delayed—if it happens at all. Normally, the next cycle would be expected to start roughly around 2020. The combined data indicate that we may soon be headed into what's known as a grand minimum, a period of unusually low solar activity.

The predicted solar "sleep" is being compared to the last grand minimum on record, which occurred between 1645 and 1715. Known as the Maunder Mini-mum, the roughly 70-year period coincided with the coldest spell of the Little Ice Age, when European canals regularly froze solid and Alpine glaciers encroached on mountain villages.

TRUTH:

ABOUT A MILLION EARTHS COULD FIT INSIDE THE SUN.

Headed for a Cold Snap?

"We have some interesting hints that solar activity is associated with climate, but we don't understand the association," said Dean Pesnell, project scientist for NASA's Solar Dynamics Observatory (SDO). Also, even if there is a climate link, Pesnell doesn't think another grand minimum is likely to trigger a cold snap.

"With what's happening in current times—we've added considerable amounts of carbon dioxide and methane and other greenhouse gases to the atmosphere," said Pesnell, who wasn't involved in the suite of new sun stud-ies. "I don't think you'd see the same cooling effects today if the sun went into another Maunder Minimum–type behavior."

Sunspots Losing Strength

Sunspots are cool, dark blemishes visible on the sun's surface that indicate regions of intense magnetic activity. For centuries, scientists have been using sunspots—some of which can be wider than Earth—to track the sun's magnetic highs and lows. For instance, 17th-century astronomers Galileo Galilei and Giovanni Cassini separately tracked sunspots and noticed a lack of activity during the Maunder Minimum.

In the 1800s, scientists recognized that sunspots come and go on a regular cycle that lasts about 11 years. We're now in Solar Cycle 24, heading for a maximum in the sun's activity sometime in 2013.

Recently, the National Solar Observatory's Matt Penn and colleagues analyzed more than 13 years of sunspot data collected at the McMath-Pierce Telescope at Kitt Peak, Arizona. They noticed a long-term trend of sunspot weakening, and if the trend continues, the sun's magnetic field won't be strong enough to produce sunspots during Solar Cycle 25, Penn and colleagues predict.

> **TRUTH:**
> **A VOLCANIC ERUPTION IN 1883 MADE THE SUN LOOK GREEN.**

"The dark spots are getting brighter," Penn said. Based on their data, the team predicts that, by the time it's over, the current solar cycle will have been "half as strong as Cycle 23, and the next cycle may have no sunspots at all."

Sluggish Jet Streams

Separately, the National Solar Observatory's Frank Hill and colleagues have been monitoring solar cycles via a technique called helioseismology. This method uses surface vibrations caused by acoustic waves inside the star to map interior structure.

Specifically, Hill and colleagues have been tracking buried "jet streams," called torsional oscillations, encircling the sun. These bands of flowing material first appear near the sun's poles and migrate toward the equator. The bands are thought to play a role in generating the sun's magnetic field.

Sunspots tend to occur along the pathways of these subsurface bands, and the sun generally becomes more active as the bands near its equator, so they act as good indicators for the timing of solar cycles.

"The torsional oscillation . . . pattern for Solar Cycle 24 first appeared in 1997," Hill said. "That means the flow for Cycle 25 should have appeared in 2008 or 2009, but it has not shown up yet." According to Hill, their data suggest that the start of Solar Cycle 25 may be delayed until 2022—about two years late—or the cycle may simply not happen.

Crawl to the Poles

Adding to the evidence, Richard Altrock, manager of the U.S. Air Force's coronal research program for the National Solar Observatory (NSO), has observed telltale changes in a magnetic phenomenon in the sun's corona—its faint upper atmosphere.

Known as the "rush to the poles," the rapid poleward movement of magnetic features in the corona has been linked to an increase in sunspot activity, with a solar cycle hitting its maximum around the time the features reach about 76 degrees latitude north and south of the sun's equator.

The rush to the poles is also linked to the sun "sweeping away" the magnetic field associated with a given solar cycle, making way for a new magnetic field and a new round of sunspot activity.

This time, however, the rush to the poles is more of a crawl, which means we could be headed toward a very weak solar maximum in 2013—and it may delay or even prevent the start of the next solar cycle.

Quiet Sun Exciting for Science

Taken together, the three lines of evidence strongly hint that Solar Cycle 25 may be a bust, the scientists reported during a 2011 meeting of the American Astronomical Society in Las Cruces, New Mexico. But a solar lull is no cause for alarm, NSO's Hill said: "It's happened before, and life seems to go on. I'm not concerned but excited."

In many ways a lack of magnetic activity is a boon for science. Strong solar storms can emit blasts of charged particles that interfere with radio communications, disrupt power grids, and can even put excess drag on orbiting satellites. "Drag is important for people like me at NASA," SDO's Pesnell said, "because we like to keep our satellites in space."

What's more, a decrease in sunspots doesn't necessarily mean a drop in other solar features such as prominences, which can produce aurora-triggering coronal mass ejections. In fact, records show that auroras continued to appear on a regular basis even during the Maunder Minimum, Pesnell said.

Instead, he said, the unusual changes to the sun's activity cycles offer an unprecedented opportunity for scientists to test theories about how the sun makes and destroys its magnetic field. "Right now we have so many sun-watching satellites and advanced ground-based observatories ready to spring into action," Pesnell said. "If the sun is going to do something different, this is a great time for it to happen." ∎

Seeing Spots

Sunspots have been observed from Earth for centuries. Chinese astronomers first recorded seeing them with the naked eye more than 2,800 years ago. In the early 1600s, astronomers Thomas Harriot and Galileo Galilei were the first to observe them through a new invention, the telescope.

Star Caught Eating Another Star

X-Ray Flare Shows

Star light. Star bright. First star I eat tonight . . .

A tiny cannibal star has been caught munching on another star, thanks to a superbright flash of x-rays spied by cosmic hunters.

Stellar Munchies

The culprit is what's known as a neutron star, the tiny but very dense corpse of a massive star that died in a supernova blast. Sitting 16,000 light-years away, this particular neutron star is normally among the faintest objects in the x-ray sky.

> **TRUTH:**
> **THERE ARE MORE STARS IN THE UNIVERSE THAN GRAINS OF SAND ON EARTH.**

But during recent observations with the European Space Agency's XMM-Newton space telescope, the star unexpectedly surged to 10,000 times its original brightness. "A companion blue supergiant star is believed to have thrown off a gigantic clump of superheated gas from its surface, [which] got attracted by the intense gravitational field of the much smaller and denser neutron star orbiting nearby," said study leader Enrico Bozzo, an astronomer with the ISDC Data Centre for Astrophysics in Geneva, Switzerland.

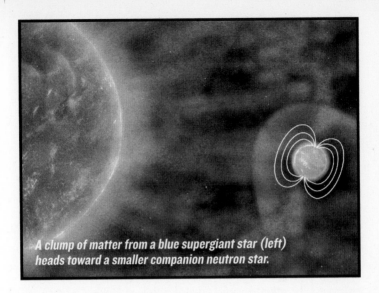

A clump of matter from a blue supergiant star (left) heads toward a smaller companion neutron star.

The lump of wayward stellar matter measured an estimated 9.9 million miles (16 million kilometers) across and took up about a hundred billion times the volume of the moon. As it became part of the neutron star, the material was heated to millions of degrees, generating a brilliant x-ray flare that lasted for four hours.

Fast Flashes

Astronomers previously knew that the neutron star and the blue supergiant are part of a stellar odd couple known as a Supergiant Fast X-ray Transient, or SFXT. These usually faint binary stars are prone to occasional flare-ups that cause them to rival the brightest x-ray sources in the sky.

Unfortunately for astronomers, these flares take place randomly only a few times a year, and they last just a couple hours, making them practically impossible to catch from beginning to end. What makes this event even harder to see is that most

> "I really could not believe this was true that we were so lucky! I didn't sleep for days. We are finally able to provide direct evidence for the existence of these clumps of matter."
> **Enrico Bozzo**
> *astronomer, describing his reaction upon discovering the cannibal star*

space-based observatories with sensitive x-ray detectors can observe only a tiny fraction of the sky at a time, and they can't be swung into action fast enough when these flares go off.

"What usually ends up happening is that these events are detected by instruments that can move very fast or have much larger fields of view, but which suffer from reduced sensitivity, so that they just can't provide a clear understanding of what caused such an event," Bozzo said.

First Proof of Star's Cannibalism

One theory was that the flares are caused by the neutron star devouring matter cast off by its hefty companion. Most massive stars generate a constant "wind" of charged particles, which pushes large quantities of stellar material in all directions into space.

Instead of a steady outflow of gas, the blue supergiant in an SFXT system may be emitting winds studded with large "bullets" of material, according to the theory. Flares happen when the neutron star gets shot by one of these clumps.

However, existing observations couldn't offer clear proof for this theory—until now. By chance, XMM-Newton caught such a flare in 2010 during a scheduled 12.5-hour observation of the SFXT system known as IGR J18410-0535.

> **TRUTH:**
> **THE BRIGHTER THE STAR, THE SHORTER ITS LIFE SPAN.**

"I really could not believe this was true that we were so lucky! I didn't sleep for days," Bozzo said. "We are finally able to provide direct evidence for the existence of these clumps of matter."

Bozzo and his team now hope to make observations of other SFXTs with XMM-Newton, to better understand the unusual flares. "We think it's the right time," he said, "to ask for an unprecedented large observational time with the space observatory and provide a final clear answer to the nature of these sources." ■

Saturn Moon

Coated in Fresh Powder

It's a winter wonderland on Saturn's moon Enceladus, where scientists have found 330 feet (100 meters) of snow on the surface.

Skiers, get your poles ready: Saturn's moon Enceladus appears to be cloaked in drifts of powdery snow, scientists announced.

Fluffy White Stuff

The researchers think superfine snowflakes are blasted out of geyser-like jets, which emanate from long fissures called tiger stripes on the moon's southern hemisphere. Some of the snow from these plumes

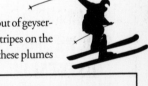

falls back to the moon's surface, coating older fractures and craters in a slow process of accumulation.

"The particles are only a fraction of a millimeter in size . . . even finer than talcum powder," study leader Paul Schenk, a planetary scientist at the Lunar and Planetary Institute in Houston, Texas, said in a statement. "This would make for the finest powder a skier could hope for."

The finding is based on new high-resolution pictures of Enceladus from NASA's Cassini orbiter, as well as global maps of color patterns that help reveal the ages of surface features.

> "The particles are only a fraction of a millimeter in size . . . even finer than talcum powder. This would make for the finest powder a skier could hope for."
> **Paul Schenk**
> *planetary scientist, Lunar and Planetary Institute, Houston*

MYSTERIOUS MOONS

1: An artist's rendering shows an **active tiger stripe on Enceladus,** including bluish regions that indicate freshly exposed water ice.

2: In 2009, Cassini captured this backlighted image that makes the 300-mile-wide (500-kilometer-wide) moon appear to be **rocket-propelled.**

3: This high-resolution Cassini picture shows a **region just north of the moon's active geysers.**

4: Cassini snapped a picture of the **E ring** in 2006.

5: A **plume of fine water-ice particles** from Enceladus stretches into space in a September 2008 Cassini picture.

1

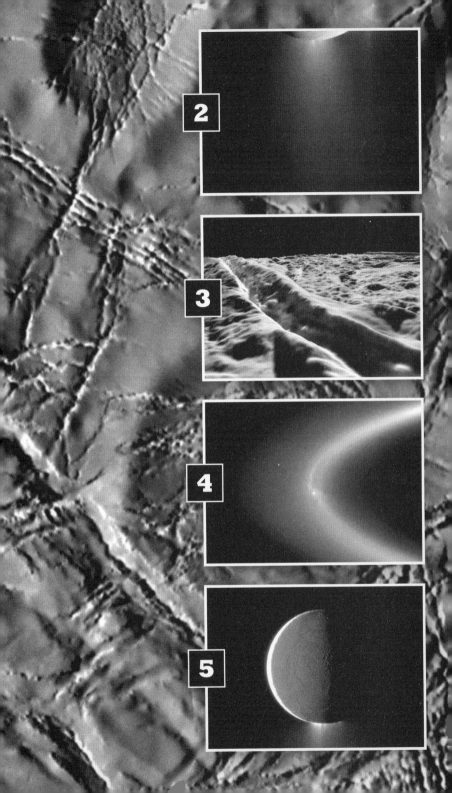

From Geyser to Blizzard

In 2005, Enceladus's icy geysers were first seen spewing from the moon's south polar region in pictures from the Cassini spacecraft. Further Cassini data have since shown that the active tiger-stripe fissures are warmer than the surrounding icy terrain, hinting that the jets are being driven by a subsurface liquid ocean. Cassini's close flybys of the moon also revealed that Enceladus's geysers may contain the chemical ingredients for life. Enceladus is so far from the sun, however, that its surface temperature is about –330 degrees Fahrenheit (–200 degrees Celsius), causing water vapor spewed from its geysers to condense into ice crystals.

> # TRUTH:
> ## SATURN HAS MORE THAN 60 MOONS.

According to the new study, as Enceladus's powdery snow falls back to the surface, it softens the contours of the underlying landscape. The rims of older craters and fissures appear to have been smoothed by the blanket of snow, while the edges of newer fractures are more distinct. Based on such images, Schenk estimates that snow has accumulated to depths of 250 to 400 feet (75 to 125 meters) in places.

Enceladus's Old Faithfuls

Scientists estimate that Enceladus's low gravity—about 1 percent that of Earth—allows some of the ice emitted by the polar geysers to jet into space rather than falling back to the moon's surface. Enough material escapes to form an entire ring of Saturn, called the E ring. The wispy E ring is so tenuous that astronomers didn't see it until 1967.

Scientists previously estimated that if Enceladus's geysers were to shut off, Saturn's E ring would dissipate within a few hundred to a few thousand years. This would mean that the entire E ring—and the geysers that feed it—could be quite recent features that we just happen to be seeing at the right time.

But the large accumulations of snow on Enceladus's surface prove otherwise, Schenk argued at a joint meeting of European and American planetary scientists. That's because the snowfall on Enceladus is incredibly light, with accumulations of less than a thousandth of a millimeter a year. For snow to build up to depths of hundreds of feet would require tens of millions of years, he said—indicating that the geysers have been active for a very long time. ∎

Star Found Shooting Water "Bullets"

How rude. Young sunlike stars have been seen spitting water out into space. Is it just a phase that all protostars go through?

Seven hundred and fifty light-years from Earth, a young, sunlike star has been found with jets that blast epic quantities of water into interstellar space, shooting out droplets that move faster than a speeding bullet.

The discovery suggests that protostars may be seeding the universe with water. These stellar embryos shoot jets of material from their north and south poles as their growth is fed by infalling dust that circles the bodies in vast disks.

"If we picture these jets as giant hoses and the water droplets as bullets, the amount shooting out equals a hundred million times the water flowing through the Amazon River every second," said Lars Kristensen, a postdoctoral astronomer at Leiden University in the Netherlands.

> **TRUTH:** COLD STARS ARE RED, WHILE HOT STARS ARE BLUE.

"We are talking about velocities reaching 200,000 kilometers [124,000 miles] per hour, which is about 80 times faster than bullets flying out of a machine gun," said Kristensen, lead author of the new study detailing the discovery, which has been accepted for publication in the journal *Astronomy & Astrophysics*.

Protostar in Perseus

Located in the northern constellation Perseus, the protostar is no more than a hundred thousand years old and remains swaddled in a large cloud—gas and dust from which the star was born.

Using an infrared instrument on the European Space Agency's Herschel Space Observatory, researchers were able to peer through the cloud and detect telltale light signatures of hydrogen and oxygen atoms—the building blocks of water—moving on and around the star.

After tracing the paths of these atoms, the team concluded that water forms on the star, where temperatures are a few thousand degrees Celsius. But once the droplets enter the outward-spewing jets of gas, 180,000-degree-Fahrenheit (100,000-degree-Celsius) temperatures blast the water back into gaseous form.

A star is born in this illustration: Gas and dust swirl inward, while polar jets spurt outward.

Once the hot gases hit the much cooler surrounding material—at about 5,000 times the distance from the sun to Earth—they decelerate, creating a shock front where the gases cool down rapidly, condense, and reform as water, Kristensen said.

Stellar Sprinkler Nourishes Galactic "Garden"

What's really exciting about the discovery is that it appears to be a stellar rite of passage, the researchers say, which may shed new light on the earliest stages of our own sun's life—and how water fits into that picture.

"We are only now beginning to understand that sunlike stars probably all undergo a very energetic phase when they are young," Kristensen said. "It's at this point in their lives when they spew out a lot of high-velocity material—part of which we now know is water."

> "If we picture these jets as giant hoses and the water droplets as bullets, the amount shooting out equals a hundred million times the water flowing through the Amazon River every second. We are talking about velocities reaching 200,000 kilometers (124,000 miles) per hour, which is about 80 times faster than bullets flying out of a machine gun."
> **Lars Kristensen**
> *postdoctoral astronomer, Leiden University, Netherlands*

Like a celestial sprinkler system, the star may be enriching the interstellar medium—thin gases that float in the voids between stars. And because the hydrogen and oxygen in water are key components of the dusty disks in which stars form, such protostar sprinklers may be encouraging the growth of further stars, the study says.

The water-jet phenomenon seen in Perseus is "probably a short-lived phase all protostars go through," Kristensen said. "But if we have enough of these sprinklers going off throughout the galaxy—this starts to become interesting on many levels." ∎

Uranus Has a Bright New Spot

Picture Shows

Uranus's cool turquoise exterior appears tranquil and serene, but astronomers have spotted a new stormy spot.

In a surprise to astronomers, Uranus recently presented onlookers with a new spot on its northern hemisphere.

Stormy Weather

Near-infrared pictures from the 8.1-meter Gemini North telescope in Hawaii have revealed a region on the giant planet that's much brighter than its surroundings. The spot is likely a tall methane cloud that reaches high enough for us to see sunlight scattered by its icy particles, said Uranus expert Heidi Hammel, executive vice president of the Association of Universities for Research in Astronomy (AURA).

> **TRUTH:**
> **THE NORTH POLE OF URANUS GETS NO SUNLIGHT FOR ABOUT 42 YEARS AT A TIME.**

The Uranian cloud is probably similar to an anvil cloud, the type of towering cumulonimbus cloud that's associated with severe thunderstorms on Earth. The cloud is also at a lower latitude on Uranus than any that have been observed before. That could mean the spot is a storm that has migrated south.

A Hubble picture of Uranus in 2005, shortly before the planet's equinox

Seeing Spots

Hammel first saw bright spots on Uranus a few years before the planet's spring equinox in 2007. She was turned on to the spots' presence thanks to a photo of Uranus in another researcher's presentation on the ice giant's moons. "I said, 'Wow, what's this?' And he said, 'I don't know, it's just the way Uranus looks.' I said, 'No it's not!'"

Subsequent observations made with the Hubble Space Telescope and ground-based telescopes revealed that the spots were most likely storms, similar to Jupiter's Great Red Spot. Storms are unusual on Uranus because the planet has very little large-scale atmospheric circulation—movements that are driven mostly by temperature differences.

That's because, unlike the other seven planets in our solar system, Uranus's axis of rotation is tilted on its side. In addition, at an average of 1.7 billion miles (2.8 billion kilometers) from the sun, the planet completes an orbit every 84 years.

During an equinox on Uranus, the planet is "completely sideways to the sun," Hammel said. Uranus's exposure to light and dark is therefore more similar to that experienced by the other planets, and the resulting temperature differences allow the icy giant's atmosphere to "turn on" and see more circulation.

The Oddly Tilted Planet

Uranus is a bit of an oddball planet, and scientists are intrigued by its curiosities:

1. Uranus spins on its side. It's the only planet with a rotational axis that's tilted almost into its orbital plane.
2. Unlike the other gas giants, Uranus doesn't emit much heat.
3. Uranus has the most powerful known winds in the solar system, blowing at more than 500 miles an hour (805 kilometers an hour).

Hubble Hopes

Hammel and other astronomers have been studying the planet's cloud activity since the last equinox to track how seasonal changes affect the weather. Until this new spot was observed, researchers had thought Uranus's spring period was over. Now astronomers aren't sure just how long the planet will continue to form such clouds.

Hammel is hoping more astronomers will study Uranus's new spot, and that enough independent confirmation of the feature will prompt Hubble managers to once again look at Uranus. After all, the last time this planet experienced a change in seasons was 1965. "This is the first opportunity in modern astronomy to look at Uranus with this detail," she said. ∎

Planets Being Pulverized

Near Giant Black Holes?

A planetary "roller derby" may be playing out around the supermassive black hole at the heart of the Milky Way, according to a new theory that could help solve a dusty mystery.

For planets and asteroids, life near a black hole can be nasty, brutish, and short. But astronomers believe that in their demise may be found the origin of giant cosmic dust clouds.

Planet Smashing

Almost all of the large galaxies observed so far have central black holes, each billions of times the mass of our sun. But about half of these cosmic monsters are obscured by dense rings of dust, and astronomers have been uncertain where all this dust comes from and how it's remained intact over time.

According to the new theory, it's possible that newly formed planets and asteroids whirling close to these giant black holes are continually being smashed to smithereens, creating the thick, donut-shaped clouds of debris.

> # TRUTH:
> ## NOTHING CAN ESCAPE FROM A BLACK HOLE.

"Near a supermassive black hole, velocities are hundreds to a few thousand kilometers per second," said study leader Sergei Nayakshin of the University of Leicester in the United Kingdom. At such speeds, "hitting an Earth-size planet with a solid object a few kilometers across could fragment [the planet] into lots of smaller fragments that, over time, become nothing more than dust."

Supermassive Attack

Supermassive black holes are the largest black holes in a galaxy, and recently, one was caught eating a star. Astronomers observed a burst of high-energy gamma rays emanating from the center of a dwarf galaxy 3.8 billion light-years away. The flash, one of the brightest and longest gamma ray bursts ever seen, was believed to be caused by a supermassive black hole destroying a star that got too close to its gravitational pull. While supermassive black holes are thought to exist in most large galaxies, such an event may happen only once every hundred million years in any given galaxy.

When Worlds Collide

In 2006, astronomers discovered a population of stars in two rotating disks around the Milky Way's central black hole. Nayakshin's team believes these stars may have also formed their own planets and asteroids, similar to the objects in our solar system. Such planets would have to be orbiting their host stars extremely closely to be able to form under the intense gravitational forces that exist near a black hole.

At the same time, stellar densities are very high near the giant black hole, so it's possible that the gravitational pull of closely passing stars frequently causes planets to unbind from their host stars and smash into each other, according to the new theory.

The dust created by these pulverized worlds might be similar to the zodiacal dust in our solar system—the result of ancient collisions between newborn planets, asteroids, and comets. A similar mechanism could be at work across the universe, filling other galaxies with dust near their central black holes.

For the study, Nayakshin and his colleagues looked at existing observations of dust clouds around supermassive black holes and compared them with computer models of dust-cloud creation in planetary systems. "If you know this process to actually work on a smaller scale in a somewhat similar setting, then chances are that it may work in the bigger system you're studying," Nayakshin said.

The team found that the type of microscopic dust generated as planets and asteroids collided would in fact block light even from actively feeding black holes, which spew intense radiation as matter falls toward the black hole, compresses, and heats up.

Galactic Protection

Of course, the future for any newly formed planet orbiting a galactic black hole may seem bleak. But the violent demise of star systems could have a protective effect for other parts of some galaxies. If the resulting clouds of dust are massive enough, Nayakshin believes, they can obscure much of the lethal x-rays and gamma rays that are constantly bellowing out from the edges of an actively feeding supermassive black hole.

While much of the area surrounding such a black hole would be sterile, Nayakshin said, "parts of the galaxy that are shielded by the ring of dust—about 50 percent of it—will have a safer and quieter environment for star and planet formation." Overall, he added, the theory raises some interesting new possibilities for the exotic environments close to supermassive black holes.

People used to think all that existed around such black holes was gas and dust, Nayakshin said. "But if we are right, then there are also planets, asteroids, and comets that exist there too—so it is a much more diverse environment than people ever thought." ∎

TRUTH: THERE IS NO TIME AT THE CENTER OF A BLACK HOLE.

Youngest Planet Picture

Gas Giant Seen in Throes of Creation

Everyone loves to show off baby pictures, and astronomers are no exception. When they got a snapshot of the youngest planet, they were proud to share it with the world.

A new picture of a Jupiter-like world swaddled in gas and dust is a direct image of what may be the youngest planet yet seen, astronomers report. The newborn gas giant, dubbed LkCa 15b, orbits a sunlike star 450 light-years away in the northern constellation Taurus. The planet orbits inside a disk of material around the star that's no more than two million years old.

The big baby planet may be up to six times the mass of Jupiter, according to theory-based calculations, and it appears to orbit 11 times farther from its parent star than Earth does from our sun. The new picture was made in near-infrared light, but "the planet would probably appear a deep red to our eye, since it's still glowing from the heat of being formed," said Adam Kraus, lead study author and an astronomer at the University of Hawaii.

> **TRUTH:**
> **MORE THAN 500 CONFIRMED PLANETS HAVE BEEN DISCOVERED ORBITING OTHER STARS.**

Mind the Gap

Kraus and colleagues zeroed in on the young star based on previous observations that showed a conspicuous gap in the star's surrounding debris disk. Such gaps are thought to be telltale signs that massive, newly formed planets are circling inside the disks—a protoplanet's gravity would clear away a wide swath of gas and dust as it accumulates matter.

"This [gap] is a huge benefit for astronomers who want to find planets—we know a planet is probably there, and we even know approximately where to look," Kraus said. "We just needed to find a way to distinguish the very faint planet from its very bright parent star."

For this, the team turned to the Keck II 10-meter telescope on the summit of Hawaii's Mauna Kea. First off, the telescope's deformable mirror was able to correct for distortions in the collected starlight caused by Earth's atmosphere. The team then used a small mask with several holes placed over the light-collecting mirrors, a method called aperture mask interferometry. This technique allowed the team to block out the light from the host star while capturing the fainter glow of the disk and its embedded planet.

Theories of Planet Formation Debunked!

Just when scientists thought they had planets all figured out, new discoveries pop up that change their thinking about just how planets behave. Here are some theories that are being overturned:

THEORY 1: All planetary orbits are roughly circular. In fact, only about one in three of the known exoplanets has a circular or near-circular orbit.

THEORY 2: With minor exceptions, everything in a star system orbits in the same plane and in the same direction. One in three exoplanets' orbits are "misaligned." Some orbit in the opposite directions of their stars' rotations, and others are tilted well out of the ecliptic.

THEORY 3: Giants the size of Neptune are rare. The size range where theory suggested there should be the fewest planets—3 to 15 times the size of Earth—has been the most commonly found.

"Baby" Pictures

Kraus and his team plan to continue observing LkCa 15b so they can pin down its temperature and orbital characteristics, such as the shape and orientation of its path around the star. The team also hopes to expand the search to other stars that have surrounding disks with gaps—and perhaps begin to answer some basic questions about early planet formation.

"We'd been looking for this kind of planet for several years, specifically because we know that observing planet formation in action would tell us a lot about how it actually works," Kraus said. "My first reaction was that this is finally going to tell us how planets really form!" ∎

"Diamond" Planet Found

May Be Stripped Star

An exotic planet as dense as diamond has been found in the Milky Way. Astronomers think the world is a former star that got transformed by its orbital partner.

> "In terms of what it would look like, I don't know I could even speculate. I don't imagine that a picture of a very shiny object is what we're looking at here."
>
> **Ben Stappers**
> *University of Manchester, on the "diamond" planet*

An odd planet was discovered orbiting what's known as a millisecond pulsar—a tiny, fast-spinning corpse of a massive star that died in a supernova. Astronomers estimate that the newfound planet is 34,175 miles (55,000 kilometers) across, or about five times Earth's diameter.

In addition, "We are very confident it has a density about 18 times that of water," said study leader Matthew Bailes, an astronomer at the Swinburne Centre for Astrophysics & Supercomputing in Melbourne, Australia. "This means it can't be made of gases like hydrogen and helium like most stars but [must be made of] heavier elements like carbon and oxygen, making it most likely crystalline in nature, like a diamond."

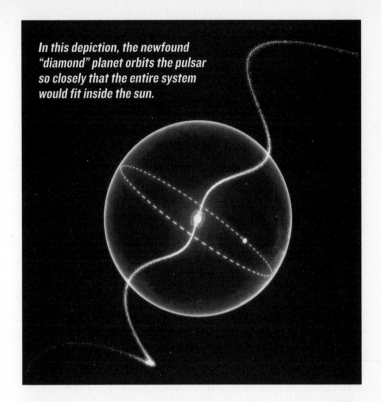

In this depiction, the newfound "diamond" planet orbits the pulsar so closely that the entire system would fit inside the sun.

Millisecond Pulsars

The new millisecond pulsar, dubbed PSR J1719-1438, lies about 4,000 light-years away in the southern constellation Serpens. Bailes and his team found the star during a pulsar survey using the radio telescope at Australia's Parkes Observatory. A pulsar is a type of stellar corpse that emits powerful beams of radio waves from its poles. If these beams sweep across Earth's field of view as the star rotates, radio telescopes on Earth can detect the star's regular pulses.

A millisecond pulsar is thought to form when the pulsar is siphoning material from a companion star. The action of eating matter speeds up the pulsar's spin to hundreds of rotations a second. So far it seems millisecond pulsars are rare, with only about a hundred found in the last 30 years. The study team found PSR J1719-1438 by using supercomputers to comb through almost 200,000 gigabytes of data—enough to fill more than 23,500 standard DVDs.

The data show that the pulsar spins more than 10,000 times a minute. The astronomers also noticed that the star's radio pulses have an unusual modulation, which the team concluded must be due to the gravitational pull of a small orbiting object.

A Stripped Star?

About 70 percent of the known millisecond pulsars have orbital companions, but PSR J1719-1438 is only the second thought to have a planetary partner. That's probably because planets don't form around millisecond pulsars in the usual way, Bailes said.

Diamonds Are Forever

The diamond planet, dubbed J1719-1438, has a stable system, with no evidence that it will change for billions of years. "Of course," notes Michael Keith of the Australia Telescope National Facility, "this also means that it could well have been around for a long time, just waiting for us to find it. Since it's likely to last for longer than the Earth or the sun, I would say that in this case, a diamond really is forever."

Astronomers think planets are created from dusty disks of material swirling around newborn stars. As this material orbits the star, gravitational interactions cause clumps to form, and the clumps build mass as they sweep through the disk. By contrast, the new study hints that pulsars can strip material away from their companions until all that's left of the consumed star is enough mass for a planetlike object.

The newfound "diamond" planet probably formed from a white dwarf star—the core of a dead sunlike star—that was being stripped of matter by the pulsar. The leftover object likely represents just 0.1 percent of the white dwarf's original mass, Bailes said. Based on their data, the team calculates that the planet orbits the pulsar in just two hours and ten minutes at a distance of about 372,822 miles (600,000 kilometers).

More Pulsar Planets Out There

Bailes and his team would now like to know exactly how rare their discovery really is. In all likelihood, this weird method of planet production requires special circumstances that rely on the white dwarf companion having a particular mass and chemical composition.

But even if the diamond planet is a result of a perfect storm of special circumstances, there should be more such worlds out there, Bailes said.

"The most exciting aspect to me is that we've only processed a small fraction of space so far," he said. "With the new supercomputers coming online, we should be in a strong position to possibly make many more discoveries like this one." ■

Seven Supernovae

Found in Single Galaxy—A First

In a galaxy 250 million light-years from Earth, astronomers have spotted a record-breaking seven supernovae, and they were all found at the same time.

Astronomers have spotted a record-breaking seven supernovae in a prodigious galaxy known as Arp 220. This galaxy is thought to have formed from the merger of two smaller galaxies and is well known to host a very intense burst of star formation, easily seen in visible wavelengths.

The galaxy Arp 220, where the seven supernovae were found

A New Record

"As far as we know, only three supernovae in a single galaxy were found at once so far, which is already an impressive number," said study leader Fabien Batejat, a Ph.D. student at Chalmers University of Technology in Onsala, Sweden. "But we can confirm seven supernovae [in a single galaxy], thanks to a 17-year monitoring of the radio sources in Arp 220."

The unprecedented find may offer a unique cosmic laboratory for studying galaxy evolution. The new data also confirm that Arp 220 is a very efficient factory for explosive star deaths, giving scientists a glimpse of how the earliest galaxies in the universe may have behaved.

Rogue Planets

A new theory suggests that when a star dies in a violent supernova, some of its planets may survive the blast but be ejected from orbit and sent wandering the galaxy. This would offer an explanation for some of the free-roaming planets that have been found and it could mean that more exist across the Milky Way. In rare cases, some survivor planets may remain bound to supernova remnants, finding new orbits around the neutron stars or black holes left behind by the explosions.

Telescope Como Revealed Supernovae

Each of the supernovae found in Arp 220 spans less than a light-year, and at such a great distance, each radio signal covers an angle in the sky less than 0.5 milliarcseconds across, Batejat said. "To give you an idea of how small this is, this size corresponds to what you would see if you would look into a straw of about 1,500 kilometers [932 miles] long," Batejat said.

"In order to see such small objects, we would need a telescope of 10,000 kilometers [6,214 miles] across, which is a bit less than the diameter of the Earth itself. But since we can't build such gigantic telescopes, we use interferometry to simulate them." In astronomy, interferometry uses the combined power of an array of telescopes—rather than a single, huge telescope—to create high-resolution images that can probe deep into the universe.

Batejat's team used 57 of the largest radio telescopes on Earth, which are spread across two continents and five countries. The project included data from the European VLBI Network, the Very Long Baseline Array, the Green Bank Telescope, and the Arecibo Observatory.

The heart of Arp 220 is highly obscured by dust that can't be penetrated by visible wavelengths. But radio waves can travel through such a dense environment to reach telescopes on Earth.

"Something Amazing"

Ultimately the data revealed about 40 radio sources near the center of Arp 220. By watching how these sources changed over time in two different radio wavelengths, astronomers could tell that seven of the objects were stars that had exploded around the same time.

Astronomers estimate that our Milky Way galaxy sees only a single supernova

> # TRUTH:
> ### INTERFEROMETRY USES THE COMBINED POWER OF MANY TELESCOPES, AS OPPOSED TO A SINGLE, HUGE TELESCOPE, TO CREATE HIGH-RESOLUTION IMAGES THAT CAN PROBE DEEP INTO THE UNIVERSE.

every hundred years, on average, Batejat said. But the highly active Arp 220, with its dynamic cycles of star birth and death, behaves more like how young galaxies probably did more than ten billion years ago.

"We hope this might lead to interesting discoveries on how stars formed [and died] in the early universe," Batejat said. What's more, such relatively fresh supernovae "are rare, and they have short lives of a few decades maximum" before they settle into supernova remnants, he said. "So discovering seven such supernovae at once is something amazing." ∎

Meteors Delivered Gold to Baby Earth

**"There's gold in them thar meteors!"
says a new study that supports the theory
that meteors delivered gold and other metals
to Earth billions of years ago.**

Not all that glitters is gold. But Earth would have a lot less of the glittery stuff if not for a massive rain of meteors about 3.9 billion years ago, according to a new study.

Based on analysis of some of the world's oldest rocks, scientists have the first direct evidence that a cataclysmic meteor shower changed early Earth's chemical composition. The find offers support for the theory that meteors delivered gold and other precious metals to infant Earth.

Gold Sank in "Magma Ball" Earth

The presence of precious metals in Earth's mantle and crust poses a puzzle because these elements are attracted to iron. When Earth first formed roughly 4.5 billion years ago, the planet was basically a ball of magma. As the planet cooled, denser material sank toward the center, eventually producing a core made mostly of iron.

But that means any iron-loving—or siderophile—elements present in the primordial magma should have also retreated toward the core. In fact, based on the composition of meteorites thought to be akin to early Earth, our planet should have enough gold in its present-day core to cover the entire globe with a 12-foot-thick (4-meter-thick) layer of the precious metal.

"All that stuff disappeared into the core, but we still find some gold around [the surface]," said study co-author Matthias Willbold of the University of Bristol. One possible answer for where the precious metals came from is that a "firestorm" of meteors called the terminal bombardment added a veneer of material to Earth's surface some 650 million years after the planet's formation.

Chemical Clues

To find proof for this theory, Willbold and colleagues studied rock samples from the Isua Greenstone Belt in Greenland. Although the Greenstone rocks date to about 3.8 billion years ago—close to the time of the terminal bombardment— "the mantle source from which these rocks are coming is from 4.5 billion years ago," Willbold said. That means the rocks should retain chemical signatures that predate the massive meteor shower.

By comparing those ancient rocks with modern ones, the researchers found that the two samples have different tungsten isotope ratios. Tungsten-182 was produced only in the first 50 million years of the solar system. But the Greenland rocks have more tungsten-182 than tungsten-184, the version of the element more commonly found in modern rocks.

"These rocks that we found on Greenland are the only ones that show an anomalous tungsten condition," Willbold said—a sign that meteor impacts did in fact change Earth's surface composition.

In general, based on the rate of impacts during the terminal bombardment, meteors slamming into Earth may have added about half a percent of the material now in the planet's mantle, Willbold said. That may not seem like much, but it works out to about 20 billion billion tons, he added. ■

> "Our work shows that most of the precious metals on which our economies and many key industrial processes are based have been added to our planet by lucky coincidence when the Earth was hit by about 20 billion billion tonnes of asteroidal material."
> **Matthias Willbold**
> *study co-author,*
> *University of Bristol*

Darkest Planet Found

Coal-black, reflecting almost no light, this newfound world is off-the-charts dark—and the cause is a mystery, experts say.

It may be hard to imagine a planet blacker than coal, but that's what astronomers say they've discovered in our home galaxy with NASA's Kepler space telescope.

Exotic Exoplanet

Orbiting only about three million miles out from its star, the Jupiter-size gas giant planet, dubbed TrES-2b, is heated to 1,800 degrees Fahrenheit (980 degrees Celsius). Yet the apparently inky world appears to reflect almost none of the starlight that shines on it, according to a new study.

"Being less reflective than coal or even the blackest acrylic paint—this makes it by far the darkest planet ever discovered," lead study author David Kipping said. "If we could see it up close it would look like a near-black ball of gas, with a slight glowing red tinge to it—a true exotic amongst exoplanets," added Kipping, an astronomer at the Harvard-Smithsonian Center for Astrophysics in Cambridge, Massachusetts.

> "Being less reflective than coal or even the blackest acrylic paint—this makes it by far the darkest planet ever discovered."
> **David Kipping**
> *astronomer,*
> *Harvard-Smithsonian Center for Astrophysics*

NASA's Planet Detector

The Earth-orbiting Kepler spacecraft was specifically designed to find planets outside our solar system. But at such distances—TrES-2b, for instance, is 750 light-years from us—it's not as simple as snapping pictures of alien worlds.

Instead, Kepler—using light sensors called photometers that continuously monitor tens of thousands of stars—looks for the regular dimming of stars. Such dips in stellar brightness may indicate that a planet is transiting (passing in front of a star, relative to Earth), blocking some of the star's light; in the case of the coal-black planet, blocking surprisingly little of that light.

Black Planet Spurs
Dimmest of Dimming

When a planet passes in front of its star, the world's shaded side faces Kepler. But as the planet begins orbiting to the side of and "behind" its star, its star-facing side comes to face the viewer. The amount of starlight grows until the planet, becoming invisible to Kepler, passes fully behind its star.

Watching TrES-2b and its star, Kepler detected only the slightest such dimming and brightening, though enough to ascertain that a Jupiter-size gas giant was the cause. The light reflected by the newfound extrasolar planet,

Experts believe the newfound gas-giant is black with a slight red glow.

or exoplanet, changed by only about 6.5 parts per million, relative to the brightness of the host star.

"This represents the smallest photometric signal we have ever detected from an exoplanet," Kipping said. What's more, as the coal-black planet passed in front of its star, the starlight's dimming was "so small that it's like the dip in brightness we would see with a fruit fly going in front of a car headlight."

The Dark Mystery of TrES-2b

Current computer models predict that hot-Jupiter planets—gas giants that orbit very close to their stars—could be only as dark as Mercury, which reflects about 10 percent of the sunlight that hits it. But TrES-2b is so dark that it reflects only 1 percent of the starlight that strikes it, suggesting that the current models may need tweaking, Kipping said.

Assuming the new study's measurements are sound, what exactly is making the new planet's atmosphere so dark? "Some have proposed that this darkness may be caused by a huge abundance of gaseous sodium and titanium oxide," Kipping said. "But more likely there is something exotic there that we have not thought of before. It's this mystery that I find so exciting about this discovery."

TrES-2b may even represent a whole new class of exoplanet—a possibility Kipping and company hope to put to the test with Kepler, which has so far detected hundreds of planets outside our solar system.

Go Kepler, Go!

The primary goal of NASA's Kepler space telescope mission is to find rocky, Earthlike planets orbiting in stars' habitable zones—the regions in which planets receive enough heat from their stars for liquid water to exist. While the finds haven't quite yet met those criteria, they do show that Kepler is working as expected, offering a "tantalizing hint at what we can expect in a few years' time," says Greg Laughlin, an astronomer at the University of California, Santa Cruz.

"As Kepler discovers more and more planets by the day, we can hopefully scan through those and work out if this is unique or if all hot Jupiters are very dark," Kipping said. Meanwhile, the very darkness of the new exoplanet suggests perhaps a catchier moniker for TrES-2b, Kipping said. "Maybe an appropriate nickname would be Erebus"—ancient Greece's god of darkness. ∎

Should Pluto Be a Planet?

Astronomers are still quarreling over Pluto's status. The debate rages on as new finds continue to fuel the arguments over whether little Pluto should regain its planetary status—or not.

Officially, Pluto is still not a planet.

But years after the ruling that demoted the icy object to dwarf planet, people continue struggling with the definition, and the debate over what exactly should be called a planet remains as contentious as any political divide. "Maybe it's just an argument over semantics, but we ought to be worried about semantics. We learned that lesson when the definition was changed," said Marc Kuchner, a planetary scientist at NASA's Goddard Space Flight Center in Maryland.

Passionate about Pluto

"After the ruling, astronomers everywhere were besieged by complaints from everyone big and small. A planet is a very personal thing—we think of the Earth, the moon, and the other planets as part of our home, and maybe that's why we got so upset about Pluto."

Since the 2006 ruling, astronomers have also made a number of scientific advances that further cloud the issue, from discoveries of planets that don't

> **TRUTH:**
> IF YOU WEIGH
> 150 POUNDS ON EARTH,
> YOU WOULD WEIGH
> ABOUT 10 POUNDS
> ON PLUTO.

orbit stars to new models of how our own solar system may have rearranged itself since birth.

Pluto's Little Oddities

The issue of whether Pluto should be a planet first arose in the 1970s, when scientists were able to refine their estimates for the mass and size of the distant body. With each new measurement, Pluto got lighter and tinier, until astronomers realized that the object is in fact smaller than Earth's moon and has a very low density.

Then in 1978, scientists announced they'd found a moon of Pluto—but one that's almost half its size, making it the largest moon in relation to its parent body. During the ensuing decades, scientists continued to find similarly large objects in Pluto's neighborhood, a region of the solar system beyond the orbit of Neptune called the Kuiper belt.

> "You don't know how you're supposed to feel about it at first. I'd like us all to think about the dwarf planets out there as new siblings that we have to get to know and learn to love."
> **Marc Kuchner**
> *planetary scientist, NASA's Goddard Space Flight Center*

Definition Decision

The biggest challenge for Pluto came in 2005, when Caltech astronomer Mike Brown announced that he'd found a Kuiper belt object more massive than Pluto—a potential tenth planet provisionally called 2003 UB313. The discovery prompted the IAU to convene a committee to decide on an official definition of a planet.

"It was a bureaucratic problem, as it had to do with naming rights for these kinds of things," said Owen Gingerich, the Harvard-Smithsonian astronomer who chaired the committee. After all, if 2003 UB313 really was a new planet, it would need a proper name on which everyone could agree.

The committee initially proposed that there be two categories of planets: the classical planets and the group of planetlike bodies beyond Neptune, to be called plutons, "as a way of tipping our hat to Pluto," Gingerich said. The planetlike object Ceres would have to be in a separate class because it

An artist's depiction of Pluto and its largest moon, Charon, are seen from one of the dwarf planet's smaller moons.

resides in the main asteroid belt, between the orbits of Mars and Jupiter. So the committee suggested it be called a dwarf planet.

The draft definition was put to a vote in 2006 at the IAU general assembly in Prague, the Czech Republic. What emerged from the session is that, to be a planet, an object must

1. be in orbit around the sun,
2. have sufficient mass for its self-gravity to overcome rigid body forces so that it assumes a hydrostatic equilibrium (nearly round) shape, and
3. have cleared the neighborhood around its orbit.

Instead of plutons, the IAU members present voted that Pluto, Ceres, and 2003 UB313—now known as Eris—would all be called dwarf planets, and that this term is not for a subclass of planets but is for a unique category of solar system objects.

Exoplanet Exceptions

At the time of the ruling, the IAU noted that the new definition does not apply to anything outside the solar system, leaving it unclear how the organization defines the planetary objects found orbiting other stars.

Since 2006, there's been an explosion in the number of these extrasolar planets, or exoplanets, known to exist, with the current count at more than 700 and rising. Many are bigger than the gas giant Jupiter, but astronomers have found an increasing number of worlds close to Earth's mass, some of

Search for Planet X

A self-taught astronomer from Kansas, Clyde Tombaugh built powerful homemade telescopes in the 1920s to scan the night skies. Hoping to get some feedback, he sent his findings to astronomers at the Lowell Observatory. Instead of a critique, he got a job offer and joined the staff at Lowell as part of their team searching for "Planet X," the elusive planet beyond Neptune. After ten months of observations and photographing 65 percent of the sky, Tombaugh's persistence paid off when he discovered the body we now know as Pluto.

which may be habitable. And in the past few years astronomers have even found rocky planets akin to Earth's mass that don't orbit stars at all.

By the current IAU definition, none of these objects are official planets because they violate the first rule about orbiting the sun. "I was disappointed when I learned that exoplanets were not included in the definition," said NASA's Kuchner.

The second part of the definition, that planets must be massive enough to be nearly round, helped draw a line between bodies such as Pluto and large asteroids such as 433 Eros, a 21-mile-long (34-kilometer-long) space rock shaped somewhat like a peanut; the third rule, whether an object has cleared its orbital neighborhood, has proved the most controversial.

Judgment Calls

Kuchner, who was a graduate student under Caltech's Brown, thinks this part of the definition is the most subjective. In baseball, he said, "If you have a foul ball, it's because the ball landed on one side of the line—that's pretty clear. But it's harder to say if something's a strike . . . That relies on someone calling it." In the case of defining a planet, IAU made the call, and "now we have to use this definition and try to play the game."

Overall, he added, the ruling was crucial for limiting the number of things in the solar system that deserve to be called planets. "We really didn't have a choice," he said. "It was either going to be eight planets or a whole lotta planets. Nature sort of forced our hand."

But other astronomers aren't in favor of placing those kinds of limits. "There are more than 200 bones in the human body. Does that mean we should redefine bones to make life easier for medical students?" argued Timothy Spahr, head of the IAU's Minor Planet Center based in Cambridge, Massachusetts.

Instead the IAU definition makes life more complicated for astronomers, he said, because the notion of whether an orbital neighborhood has been cleared remains hazy. As an example, Spahr points to the increased number of known Earth-crossing asteroids, including roughly 8,000 that are considered near-Earth objects. While these space rocks don't exactly share our planet's orbit, they do cross it, in the sense that when they are closest to the sun, they are inside Earth's orbital path.

"There's certainly no big donut where Earth is, just a big mass of objects" that could be said to share our neighborhood, he said. This mass of objects will probably always exist, as asteroids in the main belt collide, break apart, and send new material on orbits closer to Earth's. "In 50 million years our orbital neighborhood will look pretty much the same." By some counts, that means Earth will not ever clear its orbit of debris.

What Kind of Planet Are You?

For his part, Spahr favors a simpler definition than the current version. "Orbiting a star and round is a good way to start," he said. Planetary scientist Alan Stern at the Southwest Research Institute calls it the *Star Trek* criteria. He added: "If you can look out the viewfinder of the *Enterprise* and see it's round, it's a planet."

From there, Spahr said, planets could be grouped into subclasses: terrestrial planets like Earth, gas giants like Jupiter, and icy outer planets like Pluto. "We could even have a category for rogue planets, to account for the worlds that don't orbit stars."

While NASA's Kuchner thinks the current definition should stand for now, he says he's "happy that we are constantly updating our definitions and revising our vision of universe—that's what science is all about."

And of course, no matter what you call it, many astronomers will continue to see Pluto as one of the most fascinating objects in the solar system, the Minor Planet Center's Spahr said. The IAU definition "doesn't change the fact that we're going to visit Pluto with [the New Horizons] spacecraft, and scientists are still going to go hog wild over all the data we collect." ∎

TRUTH: AN 11-YEAR-OLD GIRL NAMED THE DWARF PLANET PLUTO.

"Vampire" Stars

Found in the Heart of Our Galaxy—A First

To help maintain a youthful appearance, stars may steal energy from other stars, draining their energy away.

The stellar version of vampires—stars that drain life away from other stars—has been discovered for the first time in the heart of our Milky Way galaxy.

Blue Stragglers

Called blue stragglers, these cannibal stars have been spotted in other parts of the Milky Way. They seem to lag in age next to the other stars with which they formed—appearing hotter, and thus younger and bluer. Astronomers suspect blue stragglers look so youthful because they've stolen hydrogen fuel from other stars, perhaps after colliding into their victims.

These cannibal stars are routinely found in dense star clusters, where stars have many chances to feed off each other. Now, however, scientists have found blue stragglers in the Milky Way's galactic bulge, a dense region of stars and gas surrounding the galaxy's center.

"For a long time, it was suspected there were blue stragglers in the bulge,

> **TRUTH:**
> STELLAR VAMPIRISM HAS BEEN GOING ON FOR 500,000 YEARS, AND SHOULD CONTINUE FOR 200,000 MORE.

but no one knew how many there might be," said Will Clarkson, an astronomer at Indiana University Bloomington and the University of California, Los Angeles. "At long last, we've shown they're there."

Milky Way Vampires Formed Differently?

Using NASA's Hubble Space Telescope, astronomers looked at 180,000 stars in and near the bulge. The team discovered 42 unusually blue stars that appeared much younger than the other stars.

From these 42 stars, researchers estimate that 18 to 37 of them are likely real blue stragglers that are about 10 billion to 11 billion years old. The remainder may be genuinely young stars in the bulge, or stars not actually in the bulge.

It's also possible the blue stragglers did not form by slamming into other stars and absorbing extra hydrogen fuel, as occurs in other parts of the universe. Instead, the blue stragglers in the galactic bulge may have formed by ripping hydrogen off their companion stars. This possibly occurred either when one star fed off its partner in a two-star system, or perhaps after gravitational interactions in a triple-star system had caused two of its members to merge into one.

> "We think we have a good understanding of stellar evolution, but it doesn't predict blue stragglers . . . now we have the detailed observations needed to identify how they were created. I've always enjoyed trying to get to the bottom of a mystery."
> **Aaron M. Geller**
> astronomer, Northwestern University

"There's still a lot we don't know about the details of how blue stragglers form," Clarkson said. "Finding them in the bulge provides another set of constraints that can help refine models of their formation." ■

CHAPTER 8

Human History

There's a saying: "The past is a foreign country: They do things differently there." When archaeologists and historians travel to that "country," often they're lucky enough to bring back some strange souvenirs: 120 Roman shoes from a British dig site, the location of a lost Saharan fortress, or a sword recovered from the wreck of Blackbeard's flagship. Each object—big or small—tells us something about the past, the people who lived in it, and just how weird their world was.

Oldest Known Mattress Found

The world's oldest known mattress has been unearthed in South Africa, archaeologists have announced. About the size of a double bed, it could sleep a whole family.

Thousands of years before adjustable beds and foam mattresses, a group of cave dwellers in South Africa slept on a mattress made of reeds and rushes—the world's original organic bedding. Archaeologists announced this week that they had discovered such a mattress, believed to be the world's oldest, at the Sibudu Cave site in KwaZulu-Natal.

> "There were no rules for separate eating, working, or sleeping places. Breakfast in bed may have been an almost daily occurrence."
>
> **Lyn Wadley**
> *archaeologist,*
> *University of Witwatersrand*

Original Organic Bedding

The mattress—which consists of layers of reeds and rushes—was discovered at the bottom of a pile of bedding made from compacted grasses and leafy plants. The bedding had accumulated over a period of 39,000 years, with the oldest mats dating to 77,000 years ago.

"What we have is evidence of plant bedding that is 50,000 years older than any previous site anywhere in the world," said study leader Lyn Wadley of the University of Witwatersrand in Johannesburg. The compacted layers of fossil plants—excavated from sediments 9.8 feet (3 meters) deep—show that the bedding was periodically burned, possibly to limit pests and garbage.

Insect-Repelling "Top Sheet"

What's more, researchers believe the ancient people added a "top sheet" to the bedding made of insect-repelling greenery, possibly to ward off biting bugs such as mosquitoes and flies. This fine covering of leaves may also represent the earliest known use of medicinal plants by humans.

The leaves are from the tree *Cryptocarya woodii*, or river wild-quince, a medicinal plant that produces insect-killing chemicals. While there's no evidence that the cave dwellers suffered from bed bugs, they likely used the leaves to counteract body lice, Wadley said.

Fit for the Whole Family

At an estimated 12 inches (30 centimeters) or so high, the mattresses would've been a "very comfortable" and "quite long-lasting form of bedding," Wadley said. Measuring up to 22 square feet (2 square meters), the beds were also large enough to accommodate a whole family.

For modern hunter-gatherers, such as the Inuit and Kalahari Bushmen, "the idea of just one or two people sleeping on a bed is unknown," she noted. "Hunter-gatherers tend to live with each other in kinship groups," said Wadley, whose study appeared in the journal *Science*. "It was probably the same in the Stone Age—parents, children, grannies, and all sorts of people using the same bed." ■

TRUTH: INDIGENOUS GROUPS IN AFRICA STILL USE RIVER WILD-QUINCE LEAVES TODAY TO REPEL INSECTS.

Dead Sea Scrolls Mystery Solved?

The recent decoding of a cryptic cup, the excavation of ancient Jerusalem tunnels, and other archaeological detective work may help solve one of the great biblical mysteries: Who wrote the Dead Sea Scrolls?

New clues hint that the Dead Sea Scrolls, which include some of the oldest known biblical documents, may have been the textual treasures of several groups, hidden away during wartime—and may even be "the great treasure from the Jerusalem Temple," which held the Ark of the

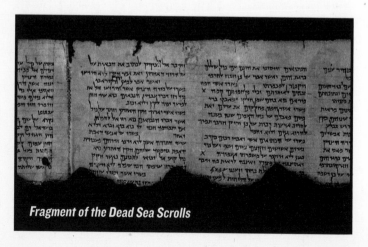

Fragment of the Dead Sea Scrolls

Covenant, according to the Bible. New theories have been generated, but the controversy over the identity of the scrolls' authors remains.

The Dead Sea Scrolls

The Dead Sea Scrolls were discovered more than 60 years ago in seaside caves near an ancient settlement called Qumran. The conventional wisdom is that a breakaway Jewish sect called the Essenes—thought to have occupied Qumran during the first centuries B.C. and A.D.—wrote all the parchment and papyrus scrolls.

TRUTH:
THE DEAD SEA IS SEVEN TIMES SALTIER THAN THE OCEAN.

But new research suggests many of the Dead Sea Scrolls originated elsewhere and were written by multiple Jewish groups, some fleeing the circa-A.D. 70 Roman siege that destroyed the legendary Temple in Jerusalem. "Jews wrote the Scrolls, but it may not have been just one specific group. It could have been groups of different Jews," said archaeologist Robert Cargill of the University of California, Los Angeles (UCLA).

The new view is by no means the consensus, however, among Dead Sea Scrolls scholars. "I have a feeling it's going to be very disputed," said Lawrence Schiffman, a professor of Hebrew and Judaic Studies at New York University (NYU).

Ritual Bathers

In 1953, a French archaeologist and Catholic priest named Roland de Vaux led an international team to study the mostly Hebrew scrolls, which a Bedouin shepherd had discovered in 1947. De Vaux concluded that the scrolls' authors had lived in Qumran, because the 11 scroll caves are close to the site.

Ancient Jewish historians had noted the presence of Essenes in the Dead Sea region, and de Vaux argued Qumran was one of their communities after his team uncovered numerous remains of pools that he believed to be Jewish ritual baths. His theory appeared to be supported by the Dead Sea Scrolls themselves, some of which contained guidelines for communal living that matched ancient descriptions of Essene customs.

"The scrolls describe communal dining and ritual bathing instructions consistent with Qumran's archaeology," explained Cargill.

Temple Treasure?

Recent findings by Yuval Peleg, an archaeologist who has excavated Qumran for 16 years, are challenging long-held notions of who wrote the Dead Sea Scrolls. Artifacts discovered by Peleg's team during their excavations suggest Qumran once served as an ancient pottery factory. The supposed baths may have actually been pools to capture and separate clay.

A Stone's Throw

In 1947, a goat wandered into a cave near the site of the ancient settlement of Qumran. A Bedouin shepherd hurled a stone into it, and the clink he heard against a pot led him to investigate. He came back out with the first of what would eventually be about 15,000 fragments of 850 scrolls hidden in the many caves along the cliffs of the Dead Sea.

And on Jerusalem's Mount Zion, archaeologists recently discovered and deciphered a 2,000-year-old cup with the phrase "Lord, I have returned" inscribed on its sides in a cryptic code similar to one used in some of the Dead Sea Scrolls. To some experts, the code suggests that religious leaders from Jerusalem authored at least some of the scrolls.

"Priests may have used cryptic texts to encode certain texts from non-priestly readers," Cargill told National Geographic News. According to an emerging theory, the Essenes may have actually been Jerusalem Temple priests who went into self-imposed exile in the second century B.C., after kings unlawfully assumed the role of high priest.

This group of rebel priests may have escaped to Qumran to worship God in their own way. While there, they may have written some of the texts that would come to be known as the Dead Sea Scrolls. The Essenes may not have abandoned all of their old ways at Qumran, however, and writing in code may have been one of the practices they preserved.

It's possible too that some of the scrolls weren't written at Qumran but were instead spirited away from the Temple for safekeeping, Cargill said. "I think it dramatically changes our understanding of the Dead Sea Scrolls if we see them as documents produced by priests," he said.

"Gone is the Ark of the Covenant. We're never going to find Noah's Ark, the Holy Grail. These things, we're never going to see," he added. "But we just may very well have documents from the Temple in Jerusalem. It would be the great treasure from the Jerusalem Temple."

From Far and Wide?

Many modern archaeologists such as Cargill believe the Essenes authored some, but not all, of the Dead Sea Scrolls. Recent archaeological evidence suggests disparate Jewish groups may have passed by Qumran around A.D. 70, during the Roman siege of Jerusalem, which destroyed the Temple and much of the rest of the city.

A team led by Israeli archaeologist Ronnie Reich recently discovered ancient sewers beneath Jerusalem. In those sewers they found artifacts—including pottery and coins—that they dated to the time of the siege. Some suggest that the finds indicate that the sewers may have been used as escape routes by Jews, some of whom may have been smuggling out cherished religious scrolls. Importantly, the sewers lead to the Valley of Kidron. From there it's only a short distance to the Dead Sea—and Qumran.

The jars in which the scrolls were found may provide additional evidence that the Dead Sea Scrolls are a collection of disparate sects' texts. Jan Gunneweg of Hebrew University in Jerusalem performed chemical analysis on vessel fragments from the Qumran-area caves.

Gunneweg described the testing process: "We take a piece of ceramic, we grind it, we send it to a nuclear reactor, where it's bombarded with neutrons, then we can measure the chemical fingerprint of the clay of which the pottery was made."

"Since there is no clay on Earth with the exact chemical composition—it is like DNA—you can point to a specific area and say this pottery was made here, that pottery was made over here." Gunneweg's conclusion: Only half of the pottery that held the Dead Sea Scrolls is local to Qumran.

Controversial Theory

Not everyone agrees with the idea that Dead Sea Scrolls may hail from beyond Qumran. "I don't buy it," said NYU's Schiffman, who added that the idea of the scrolls being written by multiple Jewish groups from Jerusalem has been around since the 1950s. "The Jerusalem theory has been rejected by virtually everyone in the field," he said.

> **TRUTH:** MOST OF THE DEAD SEA SCROLLS ARE MADE OF ANIMAL SKINS, SOME ARE PAPYRUS, AND ONE IS MADE OF COPPER. THEY ARE WRITTEN WITH A CARBON-BASED INK, FROM RIGHT TO LEFT.

"The notion that someone brought a bunch of scrolls together from some other location and deposited them in a cave is very, very unlikely," Schiffman added.

"The reason is that most of the [scrolls] fit a coherent theme and hang together. If the scrolls were brought from some other place, presumably by some other groups of Jews, you would expect to find items that fit the ideologies of groups that are in disagreement with [the Essenes]. And it's not there," said Schiffman, who dismisses interpretations that link some Dead Sea Scroll writings to groups such as the Zealots.

TRUTH:
THE DEAD SEA SCROLLS WERE FOUND IN 11 DIFFERENT CAVES.

UCLA's Cargill agrees with Schiffman that the Dead Sea Scrolls show "a tremendous amount of congruence of ideology, messianic expectation, interpretation of scripture, [Jewish law] interpretation, and calendrical dates. "At the same time," Cargill said, "it is difficult to explain some of the ideological diversity present within some of the scrolls if one argues that all of the scrolls were composed by a single sectarian group at Qumran."

Safekeeping

If Cargill and others are correct, it would mean that what modern scholars call the Dead Sea Scrolls are not wholly the work of isolated scribes. Instead they may be the unrecovered treasures of terrified Jews who did not—or could not—return to reclaim what they entrusted to the desert for safekeeping.

"Whoever wrote them, the scrolls were considered scripture by their owners, and much care was taken to ensure their survival," Cargill said. "Essenes or not, the Dead Sea Scrolls give us a rare glimpse into the vast diversity of Judaism—or Judaisms—in the first century." ∎

Legendary Swords' Sharpness

Strength From Nanotubes, Study Says

Studies of Damascus swords, legendary for their sharpness and strength, are revealing that the distinctive blades contain nanowires, carbon nanotubes, and other extremely small, intricate structures that might explain their powers.

Damascus swords, first made in the eighth century A.D., are renowned for their complex surface patterns and sharpness. According to legend, the blades can cut a piece of silk in half as it falls to the ground and maintain their edge after cleaving through stone, metal, or even other swords.

But since the techniques for making these swords have been lost for hundreds of years, no one is sure exactly why these swords are so exceptional.

Tiny Structures, Big Strength

Studies of the swords' molecular structure are now uncovering the tiny structures that may explain these properties. Peter Paufler, a crystallographer at Technical University in Dresden, Germany, and his colleagues used an electron

> **TRUTH:**
> METALSMITHS LOST THE FORMULA FOR FORGING DAMASCUS STEEL AT THE END OF THE 18TH CENTURY.

microscope to examine samples from a Damascus blade made in the 17th century. The team found the steel to contain rare-earth elements and evidence of nanowires in its microstructure.

In the journal *Nature,* the team reported that it has also discovered carbon nanotubes in the sword—the first nanotubes ever found in steel, Paufler says. The nanotubes, which are remarkably strong, run through the blade's softer steel, likely making it more resilient.

"It is a general principle of nature," Paufler said. "Materials that are softer, you can strengthen by including harder wires."

Secret Sword Techniques

Some of the nanowires Paufler and his team had previously found were made of an extremely hard iron-based mineral called cementite. In the new research, the team discovered that carbon nanotubes encase some cementite nanowires, protecting them. These nanotube-nanowire bundles may give the swords their special properties, Paufler says.

> "The important fact is that nanotubes were serving some very useful purpose even before they were discovered. This should inspire us to look for new practical applications of these remarkable nanostructures."
>
> **Andrei Khlobystov**
> *chemist,*
> **University of Nottingham, U.K.**

The bundles run parallel to the blade's surface and may help larger particles of cementite arrange in layers. These hard layers, which have softer steel in between, could help explain how the steel remains strong yet flexible. This combination of strength and flexibility makes the steel ideal for forging swords.

The blades were generally made from metal ingots prepared in India using special recipes, which probably put just the right amount of carbon and other impurities into the iron. By following these recipes and following specific forging techniques, "craftsmen ended up making nanotubes more than 400 years ago," Paufler and his colleagues wrote.

When these blades were nearly finished, blacksmiths would etch them with acid. This brought out the wavy light and dark lines that make Damascus swords easy to recognize. But it could also give the swords their sharpness, Paufler says. Because carbon nanotubes are resistant to acid, they would protect the nanowires, he theorizes. After etching, many of these nanostructures could stick out from the blade's edge, giving it tiny sawlike teeth.

Skeptical Smiths

The techniques for making the steel were lost around A.D. 1700. But many researchers are studying how to re-create the blades—even though metallurgical experts warn that the blades, though exceptional for their time, are far outperformed by modern steels.

While some scientists have claimed success, others dispute that the reproductions are truly the same as the originals. And many experts doubt that the new findings will clear things up. John Verhoeven, a metallurgist at Iowa State University at Ames who has worked on reproducing the Damascus sword-making techniques, is skeptical that Paufler and his colleagues have cracked the secret of Damascus blades. "I don't think that [the nanowires] are anything unusual," Verhoeven said. "I think those structures would be found in normal steels."

TRUTH: DAMASCUS SWORDS HAVE DISTINCTIVE-LOOKING BLADES THAT ARE CHARACTERIZED BY INTRICATE PATTERNS THAT RESEMBLE OIL SLICKS AND FLOWING WATER.

The Damascus sword is also an example of how unexpected nanosize structures can show up in materials—and sometimes give them surprising properties, experts say. But not all these nanoproperties are good. Asbestos, for example, comes in needlelike particles that cause severe lung disease. Break these particles into shorter pieces, and they are much less harmful. Because of nanomaterials' unpredictable behavior, several researchers asked for more studies of these materials and their potential side effects. ■

Lost City Revealed

Under Centuries of Jungle Growth

Something big is hiding in the jungle, and archaeologists say it's nearly a hundred ancient Maya buildings detected under a Guatemalan rain forest.

Hidden for centuries, the ancient Maya city of Holtun, or Head of Stone, is finally coming into focus. Three-dimensional mapping has "erased" centuries of jungle growth, revealing the rough contours of nearly a hundred buildings, according to research presented in 2011.

Electronic Archaeology

Though it's long been known to locals that something is buried in this patch of Guatemalan rain forest, it's only now that archaeologists are able to begin teasing out what exactly Head of Stone was.

Using GPS and electronic distance-measurement technology in 2010, the researchers plotted the locations and elevations of a seven-story-tall pyramid, an astronomical observatory, a ritual ball court, several stone residences, and other structures.

The Maya Denver

Some of the stone houses, said study leader Brigitte Kovacevich, may have

> **TRUTH:**
> THE MAYA OFTEN LEFT OFFERINGS, SUCH AS JADE OR CERAMICS, AT THE BASE OF THE ASTRONOMICAL OBSERVATORIES.

doubled as burial chambers for the city's early kings. "Oftentimes archaeologists are looking at the biggest pyramids or temples to find the tombs of early kings, but during this Late-Middle Preclassic period"—roughly 600 to 300 B.C.—"the king is not the center of the universe yet, so he's probably still being buried in the household," said Kovacevich, an archaeologist at Southern Methodist University in Dallas.

"That may be why so many Pre-classic kings have been missed" by archaeologists, who expected to find the rulers' burials at grand temples, she added.

The findings at Head of Stone—named for giant masks found at the site—could shed light on how "secondary" Maya centers were organized and what daily life was like for Maya living outside of the larger metropolitan areas such as Tikal, about 22 miles (35 kilometers) to the north, according to Kathryn Reese-Taylor, a Preclassic Maya specialist at Canada's University of Calgary.

Head of Stone, which has never been excavated, "was not a New York or Los Angeles, but it was definitely a Denver or Atlanta," said Reese-Taylor, who called the new mapping study "incredibly significant."

Head of Stone 101

The ancient Maya city of Holtun, which means "Head of Stone," is a modest site from the Pre-Classic period (600 B.C. to A.D. 250). The city had no more than 2,000 people at its peak, and preceded the famous large city-states and kingship culture for which the Maya are popularly known.

Buried Pyramid

From about 600 B.C. to A.D. 900, Head of Stone—which is about three-quarters of a mile (1 kilometer) long and a third of a mile (0.5 kilometer) wide—was a bustling midsize Maya center, home to about 2,000 permanent residents. But today its structures are buried under several feet of earth and plant material and are nearly invisible to the untrained eye.

Even Head of Stone's three-pointed pyramid—once one of the city's most impressive buildings—"just looks like a mountain enveloped in forest," said study leader Kovacevich, who presented the findings at a meeting of the Society for American Archaeology in Sacramento, California.

Jungle Thick as Thieves

Head of Stone is so well hidden, in fact, that archaeologists didn't learn of it until the early 1990s, and only because they were following the trails of looters who had discovered the site first—perhaps after farmers had attempted to clear the area, according to Kovacevich.

For thieves, the main attractions were massive stucco masks measuring up to 10 feet (3 meters) tall. Uncovered as looters dug tunnels into the buried city, the heads once adorned some of Head of Stone's most important buildings.

The temple, Kovacevich said, "would have had these really fabulously, elaborately painted stucco masks flanking the two sides of the stairway that represented human figures, snarling jaguars," and other forms.

During the Preclassic period, Head of Stone's important public buildings would have been painted primarily in blood reds, bright whites, and mustard yellows, the University of Calgary's Reese-Taylor said. Murals of geometric patterns or scenes from myth or daily life would have covered some of the buildings, she added.

> "Little is known about how kingship developed, how individuals grabbed political power within the society, how the state-level society evolved and why it then was followed by a mini-collapse between 100 A.D. and 250 A.D."
>
> **Brigitte Kovacevich**
> *archaeologist,*
> *Southern Methodist University*

King of Stars

During special events at Head of Stone, such as the crowning of a king or the naming of a royal heir, "there would have been a lot of people—not only the 2,000 people living at the site itself but all the people from surrounding areas as well. So, several thousand people," Reese-Taylor said.

Thick gray smoke and the smell of burning incense would have filled the air. Gazing up at the temple top through this haze, a visitor might have seen "ritual practitioners" performing dances and sacred rituals while adorned with elaborate feathered costumes and jade jewelry.

During the solstices or equinoxes, the crowds would have moved farther south and higher up in the city, surrounding the buildings that made up the astronomical observatory. "During the solstices, you would've been able to see the sun rising in line with the eastern structure, and the common people would have thought that the king was commanding the heavens," study leader Kovacevich said.

The researchers, though, are directing their gaze downward. Soon they hope to begin excavating residential structures and the observatory, as well as to possibly remove the undergrowth from the main temple.

And, by using ground-penetrating radar, they hope to bring Head of Stone into even sharper relief. By seeing through soil the way the previous mapping project saw through trees and brush, radar should reveal not just the rounded shapes of the city but also the hard outlines of the buildings themselves. ■

TRUTH:
ALMOST ALL MAYA SITES HAD A RITUAL BALL COURT.

The Seven New Wonders of the World

The Seven Wonders of the Ancient World have been joined by seven new wonders of the world, selected by popular vote and announced in 2007.

In 1999 the New7Wonders Foundation—the brainchild of Swiss filmmaker and museum curator Bernard Weber—decided to have a global contest to create a new list of man-made marvels that would join the classic Seven Wonders of the Ancient World, which were picked by Greek scholars about 2,200 years ago. The goal of the modern competition? To "protect humankind's heritage across the globe."

Yet the competition sparked controversy, drawing criticism from the United Nations' cultural organization UNESCO, which administers the World Heritage sites program. "This initiative cannot, in any significant and sustainable manner, contribute to the preservation of sites elected by [the] public," UNESCO said in a statement.

> ## Runners-Up
> Unsuccessful candidates included:
> 1. The giant statues of Easter Island in the Pacific Ocean
> 2. The Kremlin in Moscow, Russia
> 3. The Sydney Opera House in Australia

Getting Out the Votes

Regardless of other opinions, the New7Wonders competition was on and the voting process began in 2005. Nearly 200 nominations that came in from around the world were narrowed down to 21. The winners were voted for by Internet and phone, *American Idol* style, and the foundation reported that nearly a hundred million votes were counted.

The winners: "Christ the Redeemer" statue in Rio de Janeiro, the Great Wall of China, the Colosseum in Rome, Jordan's ancient city of Petra, the Inca ruins of Machu Picchu in Peru, the ancient Maya city of Chichén Itzá in Mexico, and India's Taj Mahal.

WONDER 1 **"Christ the Redeemer" statue, Rio de Janeiro**

Standing tall on Corcovado mountain, the 105-foot-tall (38-meter-tall) "Christ the Redeemer" was constructed between 1922 and 1931. The statue underwent a four-month restoration in 2010, when workers cleaned and replaced its stone exterior. Brazilian President Luiz Inácio Lula da Silva very openly campaigned for and encouraged his compatriots to vote for the mountaintop statue.

<table>
<tr><td>WONDER 2</td><td></td></tr>
</table>

WONDER 2   Great Wall of China

This newly elected world wonder was built along China's northern border throughout many centuries to keep out invading Mongol tribes. Constructed between the 5th century and the 16th

century B.C., the Great Wall is the world's longest man-made structure, stretching some 4,000 miles (6,400 kilometers). The best known section was built around 200 B.C. by the first emperor of China, Qin Shi Huang Di. When the new seven wonders were announced, the Chinese state broadcaster chose not to broadcast the event, and Chinese state heritage officials refused to endorse the competition.

WONDER 3 | The Colosseum, Rome, Italy

The only finalist from Europe to make it into the top seven—the Colosseum in Rome, Italy—once held up to 50,000 spectators who came to watch gory games involving gladiators, wild animals, and prisoners. Construction began around A.D. 70 under Emperor Vespasian. Modern sports stadiums still resemble the Colosseum's famous design. European sites that didn't make the cut include Stonehenge in the United Kingdom; the Acropolis in Athens, Greece; and the Eiffel Tower in Paris, France.

Petra, Jordan

Perched on the edge of the Arabian Desert, Petra was the capital of the Nabataean kingdom of King Aretas IV (9 B.C. to A.D. 40). Petra is famous for its many stone structures, such as a 138-foot-tall (42-meter-tall) temple with classical facades carved into rose-colored rock. The ancient city also included tunnels, water chambers, and an amphitheater, which held 4,000 people. The desert site wasn't known to the West until Swiss explorer Johann Ludwig Burckhardt came across it in 1812.

WONDER 5 — Machu Picchu, Peru

One of three successful candidates from Latin America, Machu Picchu is a 15th-century mountain settlement in the Amazon region of Peru. The ruined city is among the best known remnants of the Inca civilization, which flourished in the Andes region of western South America. The city is thought to have been abandoned following an outbreak of deadly smallpox, a disease introduced in the 1500s by invading Spanish forces. When the new seven wonders were announced, hundreds of people gathered at the remote, 7,970-foot-high (2,430-meter-high) site to celebrate Machu Picchu's new "seven wonders" status.

WONDER 6 — Chichén Itzá, Mexico

Chichén Itzá is possibly the most famous temple city of the Maya, a pre-Columbian civilization that lived in present-day Central America. It was the political and religious center of Maya civilization during the period from A.D. 750 to 1200. At the city's heart lies the Temple of Kukulkan—which rises to a height of 79 feet (24 meters). Each of its four sides has 91 steps—one step for each day of the year, with the 365th day represented by the platform on the top.

THE SEVEN WONDERS OF THE ANCIENT WORLD ARE:

1. THE GREAT PYRAMID OF GIZA, EGYPT. The Egyptian pharaoh Khufu built the Great Pyramid in about 2560 B.C. to serve as his tomb. It is the only remaining monument from the original list.

2. THE COLOSSUS OF RHODES, GREECE. The massive statue of the Greek sun god Helios took 12 years to build. Completed in 282 B.C., the Colossus was felled by an earthquake that snapped the statue off at the knees a mere 56 years later.

3. THE LIGHTHOUSE OF ALEXANDRIA, EGYPT. Built on the small island of Pharos between 285 and 247 B.C., the lighthouse stood for more than 1,500 years before earthquakes seriously damaged it in 1303 and 1323.

4. THE STATUE OF ZEUS AT OLYMPIA, GREECE. Created around 432 B.C., the 40-foot-tall (12-meter-tall) gold statue honored the original Olympic Games. Centuries later the statue was destroyed, although historians are still uncertain as to how.

| WONDER 7 | **Taj Mahal, India** |

The Taj Mahal, in Agra, India, is the spectacular mausoleum built by Muslim Mughal Emperor Shah Jahan to honor his beloved late wife, Mumtaz Mahal.

Construction began in 1632 and took about 15 years to complete. The opulent, domed mausoleum, which stands in formal walled gardens, is generally regarded as the finest example of Mughal art and architecture. It includes four minarets, each more than 13 stories tall. Shah Jahan was deposed and put under house arrest by one of his sons soon after the Taj Mahal's completion. It's said that he spent the rest of his days gazing at the Taj Mahal from a window. ■

5. THE HANGING GARDENS OF BABYLON, IRAQ. The Babylonian king Nebuchadnezzar II supposedly created the gardens around 600 B.C. to please his wife. It is believed they stood on the banks of the Euphrates River, but the exact location is unknown.

6. THE MAUSOLEUM OF HALICARNASSUS, TURKEY. The enormous tomb was built between 370 and 350 B.C. for King Mausolus of Caria, a region in the southwest of modern Turkey. The mausoleum stood intact until the early 15th century.

7. THE TEMPLE OF ARTEMIS, TURKEY. Completed around 550 B.C., the marble temple was burned down in 356 B.C. It was restored, only to be destroyed by the Goths in A.D. 262 and again by the Christians in 401 on the orders of St. John Chrysostom, then archbishop of Constantinople (Istanbul).

120 Ancient Roman Shoes
Found in U.K.

Construction of a U.K. supermarket uncovered a "substantial" find: an ancient Roman fort and the footwear of the soldiers who manned it.

Sophisticated Footwear

Footwear played an important role in the development of Roman civilization. Armies that had better shoes were able to travel farther and across rougher terrain. As the Roman Empire expanded, Roman shoemaking and vegetable tanning methods were introduced to newly acquired territories so that supplies would not have to be sent from Rome.

About 60 pairs of sandals and shoes that once belonged to Roman soldiers have been unearthed at a supermarket construction site in Camelon, Scotland, archaeologists say.

Soldiers' Shoes

The 2,000-year-old leather footwear was discovered along with Roman jewelry, coins, pottery, and animal bones at the site, which is located at the northern frontier of the Roman Empire. The cache of Roman shoes and sandals—one of the largest ever found in Scotland—was uncovered in 2011 in a ditch at the gateway to a second-century A.D. fort built along the Antonine Wall. The wall is a massive defensive barrier that the Romans built across central Scotland during their brief occupation of the region.

The find likely represents the accumulated throwaways of Roman centurions and soldiers garrisoned at the fort, said dig coordinator Martin Cook, an archaeologist with AOC Archaeology Group, an independent contractor in Britain. "I think they dumped the shoes over the side of the road leading into the fort," he said.

"Subsequently the ditch silted up with organic material, which preserved the shoes." Despite being discards, the hobnailed shoes are in relatively good condition, Cook added.

Newfound Fort One of Decade's Biggest Finds

While the new supermarket site also includes the remains of a first-century Roman fort and ancient field systems, excavations have centered on the area of the younger Antonine fort. "We've got evidence of a really substantial structure," Cook said. "You would have had a square fort with stone walls and three or four ditches around them."

Other finds include a Roman ax and spearhead, three or four brooches, French Samian ware—which is a high-prestige ceramic—glass, and standard pots, he said. "I would say it is one of the most important forts in Scotland," Cook added. "This will be one of the most important Scottish excavations in the last decade." The Romans are believed to have abandoned the Antonine Wall and retreated south toward England in about A.D. 165.

TRUTH:

THE 36-MILE-LONG ANTOINE WALL TOOK TWO YEARS TO BUILD AND WAS OCCUPIED FOR OVER 25 YEARS.

The Camelon dig team is on the lookout for evidence that could challenge this by suggesting the Romans stayed longer in the region. To date, however, the excavation seems to confirm that the Romans legged it—minus their footwear, of course. ■

Maya Collapse!

Caused by Man-Made Climate Change?

Self-induced drought and climate change may have caused the destruction of the Maya civilization, say scientists working with new satellite technology that monitors Central America's environment.

In early 2005, researchers from the Marshall Space Flight Center in Huntsville, Alabama, launched a satellite program, known as SERVIR, to help combat wildfires, improve land use, and assist with natural

disaster responses. SERVIR works with leaders in Central and South America, where climate change might deliver the hardest hits to ecosystems and biodiversity, say developers Tom Sever and Daniel Irwin. If the governments heed the warnings, the data may truly save lives, the experts add.

The researchers occasionally refer to the project as "environmental diplomacy," but now they can expand it to another discipline: archaeology.

Space Archaeology

SERVIR found traces of the Maya's hidden, possibly disastrous agricultural past—and is now using those lessons to help ensure that today's civilizations fare better in the face of modern-day climate change.

More than a hundred reasons have been proposed for the downfall of the Maya, among them hurricanes, overpopulation, disease, warfare, and peasant revolt. But Sever, NASA's only archaeologist, adds to evidence for another explanation. "Our recent research shows that another factor may have been climate change," he said during a meeting of the American Association for the Advancement of Science (AAAS) in Boston, Massachusetts, in 2008.

Secret Farms

One conventional theory has it that the Maya relied on slash-and-burn agriculture. But Sever and his colleagues say such methods couldn't have sustained a population that reached 60,000 at its peak. The researchers think the Maya also exploited seasonal wetlands called *bajos*, which make up more than 40 percent of the Petén landscape that the ancient empire called home.

In most cases, Maya cities encircled the *bajos*, so archaeologists thought the culture made no use of them. But groundbreaking satellite images show that the *bajos* harbor ancient drainage canals and long-overgrown fields. That ingenious method of agriculture may have backfired.

The data suggest that the combination of slash-and-burn agriculture and conversion of the wetlands induced local drought and turned up the thermostat. And that could have fueled many of the suspected factors that led to the Maya decline—even seemingly unrelated issues like disease and war.

Proven Success

The SERVIR researchers are now taking their theories to the people, showing tabletop-size satellite images to villagers and national leaders that reveal deforestation in some cases and still-lush landscapes in others. In

> **Shocking Stats**
>
> NASA researchers used computer simulations to re-create how deforestation could have played a role in aggravating the drought that may have contributed to the collapse of the Maya civilization. The worst- and best-case scenarios were modeled—100 percent deforestation in the Maya area and no deforestation. They found that a loss of all trees would have caused a three- to five-degree rise in temperature and a 20 to 30 percent decrease in rainfall.

one instance, the Guatemalan congress was inspired to create the Maya Biosphere Reserve, Central America's largest protected area, after viewing satellite imagery and seeing striking differences between Guatemala's forests and those that had been clear-cut to the north.

SERVIR, which is being supported in part by USAID and the World Bank, has also proved its worth in other ways since the program's headquarters was opened in Panama at the Water Center for the Humid Tropics of Latin America and the Caribbean (CATHALAC). In 2006, Panamanian President Martin Torrijos used the SERVIR office as his command post during widespread flooding—and when SERVIR technology forewarned of landslides, he paid attention. CATHALAC senior scientist Emil Cherrington has never deleted the text message the government sent out that day—a red alert about the landslides SERVIR said were imminent. Cherrington called the cooperation "inspiring."

> "The Maya are often depicted as people who lived in complete harmony with their environment. But like many other cultures before and after them, they ended up deforesting and destroying their landscape in efforts to eke out a living in hard times."
> **Robert Griffin**
> *archaeologist,*
> **Penn State University**

"It was a pretty neat example of the decision makers acting on information when it was provided," he said.

Learning From the Past

Despite these local efforts in environmental stewardship, however, Latin American countries are facing a heavy burden from worldwide climate change. Already, rains don't come as predictably to the Petén region, NASA archaeologist Sever said.

Local residents say their chicle trees are yielding fewer harvests, and clouds are forming higher and later in the day, sometimes not sending down rain at all, he pointed out.

Through SERVIR, Sever and his team are monitoring soil and plant responses to the changing conditions. They're also making maps for the ministries of environment and agriculture in several countries. And CATHALAC's Cherrington, who is from Belize, is using the information to predict how climate change will alter his home country in the future.

"Belize is really a country where biodiversity conservation is possible," he said, speaking at a AAAS meeting. Cherrington said precipitation will be disrupted most in the mountains and temperatures will increase the most on the coasts. SERVIR data are predicting that some bird and mammal species will be lost, but amphibians will be the hardest hit.

If satellite precipitation forecasts can be passed to farmers, they'll be able to make decisions about crops based on how much water they'll require, he added. The SERVIR scientists also hope to expand the space-based technology into other realms. They're looking to develop the kind of air quality index for Central America that is standard on U.S. weather reports.

And industry has already suggested applications that the SERVIR scientists didn't originally have in mind. A Panamanian company seeking to build solar panels asked recently if SERVIR could show the company where to find the best sun exposure. "It's kind of astounding," Cherrington said, "how space-based information can lead to making better decisions." ■

TRUTH:
THE LONGEST DROUGHT IN RECORDED HISTORY LASTED 400 YEARS IN THE ATACAMA DESERT IN CHILE, THE DRIEST PLACE ON EARTH.

Ancient Gem-Studded Teeth

Show Skill of Early Dentists

The glittering "grills" of some hip-hop stars aren't exactly unprecedented. Sophisticated dentistry allowed Native Americans to add bling to their teeth as far back as 2,500 years ago, a new study says.

Ancient peoples of southern North America went to "dentists"—among the earliest known—to beautify their chompers with notches, grooves, and semiprecious gems, according to a recent analysis of thousands of teeth examined from collections in Mexico's National Institute of Anthropology and History.

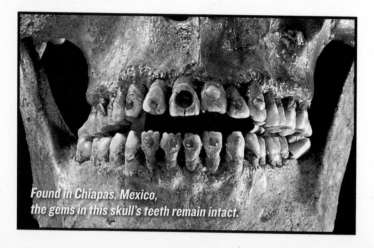

Found in Chiapas, Mexico, the gems in this skull's teeth remain intact.

Sparkling Smiles

Scientists don't know the origin of most of the teeth in the collections, which belonged to people living throughout the region, called Mesoamerica, before the Spanish conquests of the 1500s. But it's clear that people—mostly men—from nearly all walks of life opted for the look, noted José Concepción Jiménez, an anthropologist at the institute, which recently announced the findings.

TRUTH:
YOUR TEETH ARE HARDER THAN YOUR BONES.

"They were not marks of social class" but instead meant for pure decoration, he commented in an email interview conducted in Spanish. In fact, the royals of the day—such as the Red Queen, a Maya mummy found in a temple at Palenque in what is now Mexico—don't have teeth decorations, Jiménez said. Other evidence of early Mesoamerican dentistry—including a person who had received a ceremonial denture—has also been found.

Knowledgeable Dentists

The early dentists used a drill-like device with a hard stone such as obsidian, which is capable of puncturing bone. "It's possible some type of [herb-based] anesthetic was applied prior to drilling to blunt any pain," Jiménez said.

The ornamental stones—including jade—were attached with an adhesive made out of natural resins, such as plant sap, which was mixed with other chemicals and crushed bones, Jiménez said.

The dentists likely had a sophisticated knowledge of tooth anatomy, Jiménez added. For example, they knew how to drill into teeth without hitting the pulp inside, he said. "They didn't want to generate an infection or provoke the loss of a tooth or break a tooth." ∎

Mysterious Inscribed Slate

Discovered at Jamestown

An inscribed slate found in a well in Jamestown, Virginia, has presented an interesting mystery to archaeologists. Who did it belong to? What does it say? And what was it doing in a well?

Archaeologists in Jamestown, Virginia, have discovered a rare inscribed slate tablet dating back some 400 years, to the early days of America's first permanent English settlement.

Slate tablets were sometimes used in 17th-century England instead of paper, which was expensive and not reusable. According to Bly Straube, Historic Jamestown's curator, people drew games and wrote on broken roofing tiles, which could be washed off and used again and again. "Inscribed slates from this time period are rarely found in England, so little is known about them," she said.

Baby's Toy

Another recent discovery from the same Jamestown well is a brass baby's toy that's a combination whistle and teething stick. The teething-stick portion is made from coral. In the 17th century, coral was considered good for babies' gums and a magical substance that kept away evil. It may have belonged to one of the women who arrived with children in 1609.

Down the Well

Both sides of the Jamestown slate are covered with words, numbers, and etchings of people, plants, and birds that its owner likely encountered in the New World in the early 1600s. The tablet was found a few feet down in what may be the first well at James Fort, dug in early 1609 by Capt. John Smith, Jamestown's best known leader, said Bill Kelso, director of archaeology at the site.

If the well is confirmed as Smith's, it could help offer important insights into Jamestown's difficult early years. Records indicate that by 1611, the water in Smith's well had become foul and the well was then used as a trash pit. Archaeologists discovered the slate among other objects thrown into the well by the colonists.

"A Minon of the Finest Sorte"

Archaeologists and other scientists are still trying to decipher the slate, the first with extensive inscriptions to be found at any 17th-century colonial American site. The scratched and worn 5-by-8-inch (13-by-20-centimeter) tablet is inscribed with the words "A MINON OF THE FINEST SORTE." Above the words are the letters and numbers "EL NEV FSH HTLBMS 508," interspersed with symbols that have yet to be interpreted.

> "The crude drawings of birds and flora offer dramatic evidence of how captivated the English were by the natural wonders of the alien New World."
> **Bill Kelso**
> *excavation director at the Jamestown site*

"We don't know what it means yet," Kelso said. But there are some clues.

According to Straube, "minon" is a 17th-century variation of the word "minion" and has numerous meanings, including "servant," "follower," "comrade," "companion," "favorite," or someone dependent on a patron's favor. A minion is also a type of cannon—and archaeologists have found shot at the James Fort site that's the right size for a minion.

Drawings on the slate depict several different flower blossoms and birds that may include an eagle, a songbird, and an owl. "The crude drawings of birds and flora offer dramatic evidence of how captivated the English were by the natural wonders of the alien New World," excavation director Kelso said. There's also a sketch of an Englishman smoking a pipe and a man, whose right hand seems to be missing, wearing a ruffled collar.

Although the age of the tablet is not yet known, archaeological evidence—including turtle and oyster shells, Indian pots, trade beads, mirror glass, early

A sketch of a man can be seen on this slate tablet from Jamestown.

pipes, medicinal jars, and military items—indicates that it was deposited in the well during the early years of James Fort, which was established in 1607. If it's Smith's well, archaeologists believe the tablet could date to 1611, when the well was probably filled in, or earlier.

Looking for Clues

It's impossible to know yet who the slate's owner—or owners—may have been. Straube said an image that looks like a palmetto tree, normally found from South Carolina to the Caribbean, suggests that the drawings may have been made during the voyage from England to Jamestown through the West Indies, once a common route to the New World.

Or, she said, the slate could have been used by a colonist who was among about 140 castaways from the *Sea Venture* shipwreck in 1609. They were stranded in Bermuda for ten months and arrived at Jamestown in the spring of 1610.

Drawings of three rampant lions, used in the English coat of arms during the 1603 to 1625 reign of King James I, have also been discovered on the slate and could mean that the slate's owner was someone involved with government.

Archaeologist Kelso suggests the slate may have belonged to William Strachey, who served as secretary of the colony. He was among the shipwrecked colonists in Bermuda and arrived in Jamestown in 1610.

> **TRUTH:**
> JAMESTOWN WAS A POOR LOCATION FOR A SETTLEMENT BECAUSE THE DRINKING WATER WAS BAD, THE LAND WAS WET, AND THERE WERE TOO MANY DISEASE-CARRYING MOSQUITOS.

Straube, the curator, also said the tablet may have been used by someone living in Jamestown who died in the winter of 1609 to 1610, known to colonists as the Starving Time, when the fort was under siege. Only about 60 of 200 people survived.

Near the slate archaeologists have found butchered bones and teeth from horses, as well as dog bones, that may date back to the infamous winter, when colonists resorted to eating their horses and dogs to survive.

It's also possible that the tablet was used by more than one person. "There seems to be a difference in the style of handwriting," Straube noted.

Scratches and Grooves

The images on the tablet are difficult to see because they are the same dark gray color as the slate and they overlap. The colonists would have written on the tablet with a small, rectangular pencil made of slate with a sharp point. This would have made a white mark—and fortunately for archaeologists today, it also left a scratch.

"You can wipe off the mark, but you can't completely erase the groove," Kelso, the archaeologist, said. "That's why we have layer upon layer of drawings. In a way it's archaeology. If one groove cuts across another groove, you could tell which one was the most recent."

He hopes eventually to sort out the sequence of the images with the help of NASA, where scientists at the Langley Research Center are using a high-precision, three-dimensional imaging system similar to a CT scanner to help isolate the layers and provide a detailed analysis of the tablet.

Artifacts Found at Jamestown

John Smith's well has yielded other fascinating relics, some originating from far corners of the world. The artifacts include:

1. Chinese porcelain wine cups
2. Venetian glassware
3. Pieces of a Bartmann jug, a German clay vessel shaped like a bearded man
4. Telescope lenses
5. Wall tiles from Holland
6. A jet crucifix

John Smith's Well?

Determining whether this is in fact Smith's well will be key to understanding Jamestown's most difficult early years. According to colonists' accounts, water in Smith's well became brackish within a year after it had been dug. Some experts think foul water may have been a major cause of death.

Located near the James River, the well was discovered in 2008, and archaeologists believe it was dug before a well dating to 1611, which is located farther away from the river. Kelso said the colonists, having learned a difficult lesson from Smith's well, would have dug their second well as far from the river as possible, to try to avoid contamination by the brackish river water.

Archaeologists have dug down 5 feet (1.5 meters) so far, and the pit has narrowed into a more well-like, circular shape, which may reach 9 to 15 feet (2.7 to 4.5 meters) into the ground. Kelso said they won't know for sure if it's Smith's well until they get to the bottom and date the objects there.

Finding the well, he said, "will give us a chance to really look at the health issue and figure out what spoiled the water." Some clues to the mysteries contained in the 400-year-old slate might emerge then too. ■

"Lost" Fortresses of Sahara

Revealed by Satellites

Satellite photographs have exposed well-preserved settlements of a mysterious African kingdom discovered in Libya.

Real-life "castles in the sand" made by an ancient culture have been revealed in the Sahara, archaeologists say. New satellite photographs show more than a hundred fortress settlements from a "lost" civilization in southwestern Libya.

Lost Civilization Found

The communities, which date to between about A.D. 1 and 500, belonged to an advanced but mysterious people called the Garamantes, who ruled from roughly the second century B.C. to the seventh century A.D. Researchers uncovered the Garamantes' walled towns, villages, and farms after poring over modern satellite images—including high-resolution pictures used by the oil industry—as well as aerial photos taken during the 1950s and 1960s.

> **TRUTH:** DATA COLLECTED BY SATELLITES SUGGEST THAT THE SAHARA HAD A WET CLIMATE THAT SUPPORTED VAST FORESTS 12,000 YEARS AGO.

Located about 620 miles (1,000 kilometers) south of Tripoli, the fortresses were confirmed based on Garamantes pottery samples collected during an early-2011 expedition. That field trip was cut short by the civil war that would end the 42-year regime of Libyan leader Moammar Gadhafi.

The Good Life

The Garamantes certainly had a comfortable existence that was unique for a desert-dwelling people. As a result of their aggressive mentality, the acquisition of slaves, and the production of water, the Garamantes lived in planned communities and ate locally grown grapes, figs, sorghum, pulses, barley, and wheat, as well as imported goods like wine and olive oil. Says archaeologist Andrew Wilson of the University of Oxford, "The combination of their slave-acquisition activities and their mastery of foggara irrigation technology enabled the Garamantes to enjoy a standard of living far superior to that of any other ancient Saharan society."

"We were astonished to see the level of preservation" of the ancient mud-brick compounds, said project leader David Mattingly, of the U.K.'s University of Leicester.

"Although the walls of these sites have slumped a little bit, mainly due to wind erosion, they are still standing 3 to 4 meters [10 to 13 feet] high in places," he said.

Powerful African Kingdom

Archaeologists could have easily mistaken the well-planned, straight-line construction for Roman frontier forts of similar design, Mattingly observed. "But, actually, this is beyond the frontiers of the Roman Empire—these sites are markers of a powerful native African kingdom," he said.

What's more, the scientists were surprised that the sites—which include cemeteries and agricultural fields—are so tightly clustered. For instance, an area of 1.5 square miles (4 square kilometers) contained at least ten village-size settlements—"that's an extraordinary density," Mattingly said.

Previous knowledge of the Garamantes is based mainly on excavations at their capital, Jarma, some 125 miles (200 kilometers) to the northwest, as well as on ancient Roman and Greek texts. "We've built up a picture of them as being a very sophisticated, high-level civilization," Mattingly said. "They've got metallurgy, very high-quality textiles, a writing system . . . those sorts of markers that would say this is an organized, state-level society," he said.

Cash-strapped heritage authorities in Libya have been unable to conduct field research, leaving a gap in knowledge of the ancient civilization, according to University of Oxford archaeologist Philip Kenrick, who was not involved in the new research. That's why Mattingly and his team—aided by a $3.4-million grant from the E.U.'s European Research Council—have "been breaking new ground on an unprecedented scale," Kenrick said.

Creators of the Green Sahara

The newfound remains are also a testament to the Garamantes' advanced irrigation technology, which enabled them to create green oases in the desert. "It's a deep Saharan, hyper-arid environment, and it's only people's ability to exploit groundwater that can change that," project leader Mattingly said.

The Garamantes mined reservoirs of prehistoric water using underground canals to cultivate Mediterranean crops—such as wheat, barley, figs, and grapes—and sub-Saharan African sorghum, pearl millet, and cotton. Mattingly and colleagues have calculated that 77,000 man-years of labor went into constructing the underground water channels—a figure that doesn't include digging the wells or maintenance. A man-year is a unit of the work done by a person in a year.

All Dried Up?

What happened to the Garamantes remains a riddle, but Mattingly's team suspects that the desert communities declined once groundwater supplies diminished. Paul Bennett, head of mission of the U.K.-based Society of Libyan Studies, agreed that's a likely scenario. "Groundwater is a nonrenewable source—as soon as you've tapped the reservoir and emptied it, it's not going to fill again," said Bennett, who was not involved in the new research.

The collapse of the Roman Empire, and increasing conflict in the Mediterranean region, would have also seriously affected the trans-Saharan trade upon which the desert civilization depended, added Oxford's Kenrick. ∎

> "The new archaeological evidence is showing that the Garamantes were brilliant farmers, resourceful engineers, and enterprising merchants who produced a remarkable civilization."
> **David Mattingly**
> *project leader,*
> *University of Leicester, U.K.*

Spawn of Medieval "Black Death" Bug Still Roam the Earth

The Black Death was more than the deadliest plague outbreak on record: The epidemic appears to be responsible for the cases of plague that still infect humans today.

The Black Death killed millions of people in medieval Europe and continues to kill today. But another epidemic remains unlikely because recent studies show that the plague bacterium has changed little in the last 600 years.

> **TRUTH:**
> THE NURSERY RHYME "RING AROUND THE ROSY" IS TRACED TO THE PLAGUE'S ROSE-COLORED LESIONS AND DEADLY SPREAD.

Same Old Plague

The new findings are based on bacteria recovered from skeletons found in a mid-1300s cemetery for Black Death victims in London, England. The grave excavation was undertaken by the Museum of London Archaeology.

Kirsten I. Bos of McMaster University in Ontario, Canada, and Verena J. Schuenemann of the University of Tübingen in Germany led an effort to sequence the genome of the Black Death pathogen, *Yersinia pestis,* recovered from the medieval grave.

After examining *Y. pestis* samples from 46 teeth and 53 bones, the team determined that the plague hasn't changed much, genetically speaking, in more than 600 years. The result "indicates that contemporary *Y. pestis* epidemics have their origins in the medieval era," the study team writes.

One Giant Leap for Bacterium

Plague was already known to have evolved from a related soil-dwelling bacterium. The Black Death version of the pathogen has an additional segment of DNA that allows it to infect humans. Once plague made that leap, the pathogen spread wildly, carried by fleas that in turn traveled on rats—common hitchhikers on cargo ships and other trade vehicles.

Once infected with *Y. pestis,* a person can develop bubonic plague, an infection of the lymph nodes, or the rarer pneumonic plague, a secondary infection of the lungs. When the plague arrived in Europe in the 1340s, it killed about 30 to 50 million people—or up to half the continent's population—in five years.

A man bears the symptoms of the bubonic plague in this 14th-century engraving.

Modern Plague Followed Linear Evolution

Today, plague is still spread mainly by fleas on burrowing rodents. The disease hits up to 3,000 people worldwide—most commonly in the United States, Madagascar, China, India, and South America. With treatment, 85 percent of modern victims survive the plague.

The fact that the bacterial genome has been slow to change hints that modern medical knowledge and general susceptibility—not a less virulent version of *Y. pestis*—may be why the plague no longer devastates populations.

The slow change could be partly because just one strain of plague exist worldwide, and so the bacteria can evolve only in linear fashion. By contrast, influenza "can change very rapidly due to recombination between co-circulating strains, which ultimately led to the tremendously virulent 1918 strain," said study co-author Hendrik Poinar of McMaster University. The 1918 flu killed at least 50 million people—more than World War I—and was especially remarkable because it hit healthy adults rather than just the elderly and the very young.

Due to the plague's slow evolution, today's antibiotics are effective against modern *Y. pestis*—and the drugs would have been effective against the Black Death, too. "This again emphasizes the power of modern-day medicine and the antibiotic toolkit we have at our disposal," Poinar said. ■

> "Then the dreadful pestilence made its way along the coast by Southampton and reached Bristol, where almost the whole strength of the town perished, as it was surprised by sudden death; for few kept their beds more than two or three days, or even half a day."
>
> **Henry Knighton**
> *ca 1348,* Chronicon

Oldest Readable Writing

in Europe

Found at a site tied to myth, a Greek tablet with the oldest readable writing in Europe survived only by accident, experts say.

Marks on a clay tablet fragment found in Greece are the oldest known decipherable text in Europe, a new study says. Considered "magical or mysterious" in its time, the writing survives only because a trash heap caught fire some 3,500 years ago.

Ties to *The Iliad*

Found in an olive grove in what is now the village of Iklaina, the tablet was created by a Greek-speaking Mycenaean scribe between 1450 and 1350 B.C., archaeologists say. The Mycenaeans—made legendary in part by Homer's *Iliad,* which fictionalizes their war with Troy—dominated much of Greece from about 1600 to 1100 B.C.

> **TRUTH:**
> WRITING FOUND IN CHINA, EGYPT, AND MESOPOTAMIA IS BELIEVED TO DATE BACK AS FAR AS 3000 B.C.

So far, excavations at Iklaina have yielded evidence of an early Mycenaean palace, giant terrace walls, murals, and a surprisingly advanced drainage system, according to dig director Michael Cosmopoulos. But the tablet, found

Was the Trojan War Real?

Homer wrote his epic poem *The Iliad* in the eighth or ninth century B.C., several centuries after the city is supposed to have fallen. The story of the Trojan War and its great warriors and kings—Achilles and Hector, Meneleus and Priam—has captured imaginations for millennia. But was Troy real? Most scholars say yes: Troy did exist, and it did fall. Many believe that the Trojan War was not a single event, but a process that occurred over a long period of time.

in 2010, is the biggest surprise of the multiyear project, Cosmopoulos said.

"According to what we knew, that tablet should not have been there," the University of Missouri-St. Louis archaeologist told National Geographic News.

First, Mycenaean tablets weren't thought to have been created so early, he said. Second, "until now tablets had been found only in a handful of major palaces"—including the previous record holder, which was found among palace ruins in what was the city of Mycenae.

Although the Iklaina site boasted a palace during the early Mycenaean period, by the time of the tablet, the settlement had been reduced to a satellite of the city of Pylos, seat of King Nestor, a key player in *The Iliad*. "This is a rare case where archaeology meets ancient texts and Greek myths," Cosmopoulos said in a statement.

Tablet Preserved by Cooking

The markings on the tablet fragment—which is roughly 1 inch (2.5 centimeters) tall by 1.5 inches (4 centimeters) wide—are early examples of a writing system known as Linear B. Used for a very ancient form of Greek, Linear B consisted of about 87 signs, each representing one syllable.

The Mycenaeans appear to have used Linear B to record only economic matters of interest to the ruling elite. Fittingly, the markings on the front of the Iklaina tablet appear to form a verb that relates to manufacturing, the

researchers say. The back lists names alongside numbers—probably a property list.

Because these records tended to be saved for only a single fiscal year, the clay wasn't made to last, said Cosmopoulos, whose work was funded in part by the National Geographic Society's Committee for Research and Exploration.

"Those tablets were not baked, only dried in the sun and [were], therefore, very brittle... Basically someone back then threw the tablet in the pit and then burned their garbage," he said. "This fire hardened and preserved the tablet."

Not the Oldest Writing

While the Iklaina tablet is an example of the earliest writing system in Europe, other writing is much older, explained classics professor Thomas Palaima, who wasn't involved in the study, which was published in the journal *Proceedings of the Athens Archaeological Society*. For example, writings found in China, Mesopotamia, and Egypt are thought to date as far back as 3000 B.C. Linear B itself is thought to have descended from an older, still undeciphered writing system known as Linear A. And archaeologists think Linear A is related to the older hieroglyph system used by the ancient Egyptians.

Magical, Mysterious Writing

Still, the Iklaina tablet is an "extraordinary find," said Palaima, an expert in Mycenaean tablets and administration at the University of Texas-Austin. In addition to its sheer age, the artifact could provide unique insights about how ancient Greek kingdoms were organized and administered, he added.

TRUTH: THE ABILITY TO READ AND WRITE WAS VERY RESTRICTED DURING THE MYCENAEAN PERIOD AND WAS CONSIDERED MAGICAL OR MYSTERIOUS BY MOST PEOPLE.

For example, archaeologists previously thought such tablets were created and kept exclusively at major state capitals, or "palatial centers," such as Pylos and Mycenae. Found in the ruins of a second-tier town, the Iklaina tablet could indicate that literacy and bureaucracy during the late Mycenaean period were less centralized than previously thought.

Palaima added that the ability to read and write was extremely restricted during the Mycenaean period and was regarded by most people as "magical or mysterious." It would be some 400 to 600 years before the written word was demystified in Greece, as the ancient Greek alphabet overtook Linear B and eventually evolved into the 26 letters used on this page. ■

Blackbeard's Ship Yields

Ornamental Sword

Archaeologists have made a remarkable discovery among the remains of Blackbeard's flagship *Queen Anne's Revenge*. Could it be the notorious pirate's sword?

Since 1997, archaeologists have been excavating the wreck of the *Queen Anne's Revenge*, the flagship of the 18th-century pirate Blackbeard. The remains of the ship lay off the coast of North Carolina and contain a rich trove of artifacts and treasures. And now archaeologists have made a remarkable discovery: Could it be the pirate's sword?

> # TRUTH:
> ## *QUEEN ANNE'S REVENGE* WAS ORIGINALLY A FRENCH SHIP, *LE CONCORDE*, CAPTURED BY BLACKBEARD IN 1717.

Queen Anne's Revenge

Blackbeard's brief career as a pirate lasted only about two years, but during that time he became one of history's most feared outlaws. Operating in the West Indies and off the coast of colonial America, he struck terror into the hearts of commercial ships' captains and once held the entire city of Charleston, South Carolina, hostage.

After running aground on a sandbar in 1718 near the town of Beaufort, Blackbeard's ship was abandoned but likely remained intact and partly above water for as long as a year before collapsing and disintegrating, according to archaeologist David Moore of the North Carolina Department of Cultural Resources.

Blackbeard's Sword?

Because the ship remained above water for so long, "the pirates would have had ample opportunity to take anything that they thought valuable," said Moore. The newfound hilt may have been left behind because it was unwanted, or it may have been inaccessible, according to Moore's colleague Wendy Welsh, a conservator on the project. Could this partly gilded hilt have held Blackbeard's sword? There's no way to know for sure.

Elegantly Wasted

Not the Jolly Roger

According to legend, Blackbeard's distinctive pirate flag inspired fear in his enemies when they saw it flying from his ship. On a black background, the flag depicted a skeleton toasting the devil with one hand and thrusting a spear into a heart with the other.

Recovered from the wreck site in 2008, the sword's quillon (also known as a crossguard) could have been made in England or France, according to Welsh. Beyond the hilt, only a stump of the blade remains, but Welsh said Jan Piet Puype, a Dutch arms historian, thinks the weapon was probably relatively short and was carried by a gentleman with some status—at least before a pirate got hold of it.

Although it could have been used for self-defense, the sword was mainly a decorative accessory and was manufactured sometime between the mid-17th century and the early 18th century, according to Puype.

In 2010, divers recovered a carved antler believed to be the sword's handle, two years after the quillon was found. Experts hope to determine the antler's origin, which could help pinpoint where the weapon was made, Welsh said. But, she added, there's "no way of knowing" how the sword ended up aboard Blackbeard's flagship.

Archaeology in Bloom

Where others see flowers and faces, *Queen Anne's Revenge* experts see evidence on the part of the sword's handle called the pommel. For example, the flowers are irises—aka fleurs-de-lis, the royal emblem of France. Before Blackbeard captured the ship and renamed it the *Queen Anne's Revenge,* it had been a private French slave ship, *Le Concorde.* The pommel's floral embellishments may be clues that the sword too originated in France.

PIRATE BOOTY

1 : *Still underwater at the* Queen Anne's Revenge *wreck site, this* **anchor** *is about 12 feet (3.7 meters) long and 10 feet (3 meters) wide. Eventually it will be brought to the surface, archaeologists say.*

2 : *The sword hilt—found in pieces but later reassembled—was revealed to the public in 2011.*

3 : *A gilded pommel discovered in the remains of the* Queen Anne's Revenge.

4 : *Mineral encrustations cover a* **wooden pulley** *called a deadeye.*

4

Deadeye

The newly revealed, 3-foot-long (91-centimeter-long) wooden deadeye, a pulley that would have helped hold sails in place on the pirate flagship, is almost completely encrusted. The deadeye survived the ocean waters that claimed most of the rest of the *Queen Anne's Revenge* because the pulley became overgrown with minerals and was buried for almost 300 years in sand and sediment. Until they can remove the minerals and treat the artifact so it won't deteriorate in air, conservators are keeping the deadeye underwater in a lab tank.

> # TRUTH:
> ## THERE ARE NO ACCOUNTS OF BLACKBEARD EVER BURYING TREASURE.

Blackbeard's Staying Power

Divers have been removing artifacts from the wreck site since it was discovered in the mid-1990s. Only about half of the wreck site has been excavated, which should leave archaeologists with plenty to do. ∎

King Solomon's Wall Found

Proof of Bible Tale?

A 3,000-year-old defensive wall might be unprecedented archaeological support for a Bible passage on King Solomon.

A 3,000-year-old defensive wall possibly built by King Solomon has been unearthed in Jerusalem, according to the Israeli archaeologist who led the excavation. The discovery appears to validate a Bible passage, she says.

Solomon's Wall?

The tenth-century B.C. wall is 230 feet (70 meters) long and about 20 feet (6 meters) tall. It stands along what was then the edge of Jerusalem—between the Temple Mount, still Jerusalem's paramount landmark, and the ancient City of David, today an Arab neighborhood called Silwan.

Fill 'er Up

A standard bottle of wine holds 0.75 liter (about 0.20 gallon), but wine aficionados can purchase larger bottles that have special names. A "Solomon" is one of the largest. Named for the biblical king, it holds 20 liters (5.2 gallons)—26 times as much as a standard bottle of wine.

The stone barrier is part of a defensive complex that includes a gatehouse, an adjacent building, and a guard tower, which has been only partially excavated, according to Eilat Mazar, who led the dig for the Hebrew University of Jerusalem. During the years, the structures have been partially demolished—their building materials scavenged for later structures—and what remained was buried under rubble, Mazar said.

The Bible's First Book of Kings—widely believed to have been written centuries after the time period in question—says Solomon, king of Israel, built a defensive wall in Jerusalem. The new discovery is the first archaeological evidence of this structure, Mazar says.

Bearing Out a Bible Passage?

Ancient artifacts found in and around the complex pointed Mazar to the tenth-century B.C. date. "We don't have many kings during the tenth century that could have built such a structure, basically just David and Solomon," she said.

According to the Bible, King David, of David-and-Goliath fame, was the father of King Solomon, who is said to have built the First Temple of Jerusalem on the Temple Mount. Ceramics found near the wall helped narrow the date down, being of a level of sophistication common to the second half of the tenth century B.C.—King Solomon's time, according to Mazar.

Three-foot-tall (one-meter-tall) earthenware storage vessels were found near the gatehouse, one of them with a Hebrew inscription indicating the container belonged to a high-ranking government official. Figurines typical

An excavated Jerusalem site includes a wall claimed to have been built by the biblical King Solomon.

of tenth-century B.C. Jerusalem—including four-legged animals and large-breasted women likely symbolizing fertility—were also uncovered, as were jar handles bearing impressions reading "to the king" and various Hebrew names, she said.

The artifacts may hint at the area's street life in biblical times. Here ancient Jerusalemites would have gathered around the wall's city gate to trade, settle disputes via street-side judges, engage in ritual practices, and stock up on water and supplies for treks out of the city, Mazar said.

How Reliable Is the Bible?

Tel Aviv University archaeologist Israel Finkelstein, who was not involved in the excavation, agrees that it's possible King Solomon constructed the wall. But Finkelstein cautioned against leaning too heavily on the Bible to interpret the findings.

Familiarity with religious or historic texts connected to any given site is important, he said, but their usefulness can vary. "It depends upon the text. Each has its own characteristics, each needs to be approached differently," he said. "There is the question as to when it was written—300 years after, or at the time of the events? What are its goals and its ideology? Why was it written?"

For her part, Mazar believes it's natural for archaeologists excavating in the Holy Land to consult with biblical texts along with other ancient documents. "I don't believe there is an archaeologist who would excavate a site upon which texts have been written without being familiar with those texts," she said. ■

> **"I think that with archaeology, we need to use every possible source of data at our disposal. If you were interested in ancient India, you'd want to have an objective look at the *Mahabharata*. We try to create an objective historical archaeology."**
> **Thomas Levy**
> anthropologist, University of California, San Diego, on the importance of considering diverse sources of information when examining a historical site

Vikings Navigated

With Translucent Crystals?

Vikings may have navigated by looking through a type of crystal called Icelandic spar, a new study suggests.

In some Icelandic sagas—embellished stories of Viking life—sailors relied on so-called sunstones to locate the sun's position and steer their ships on cloudy days. These sunstones have turned out to be a true part of the story, a new study says.

> "The Vikings could have discovered this, simply by choosing a transparent crystal and looking through it through a small hole in a screen. The ... knowledge of the polarization of light is not necessary."
>
> **Guy Ropars**
> *physicist and study researcher,*
> *University of Rennes, France*

Polarizing Properties

Sunstones would have worked by detecting a characteristic of sunlight called polarization. Polarization is when light—which normally radiates randomly from its source—encounters something, such as a shiny surface or fog, that causes the rays to assume a particular orientation.

Due to this property, as sunlight moves through the atmosphere, the resulting polarization gives away the direction of the original source of the light. Detecting light's polarization is a natural ability of some animals, such as bees.

In 1969, a Danish archaeologist suggested real-life Vikings might have employed sunstones to detect polarized light, using the stones to supplement sundials, stars, and other navigational aids. Since then, researchers have been probing how such a sunstone might have worked. On that point, though, the sagas were silent.

Crystal Clear?

Now, Guy Ropars, a physicist at the University of Rennes in France, has conducted an experiment with a potential Viking sunstone: a piece of Icelandic spar recently found aboard the *Alderney*, a British ship that sank in 1592.

In the laboratory, Ropars and his team struck the piece of Icelandic spar with a beam of partly polarized laser light and measured how the crystal separated polarized from unpolarized light. By rotating the crystal, the team found that there's only one point on the stone where those two beams were equally strong—an angle that depends on the beam's location.

That would enable a navigator to test a crystal on a sunny day and mark the sun's location on the crystal for reference on cloudy days. On cloudy days, a navigator would only be able to use the relative brightness of the two beams.

TRUTH: ICELAND SPAR IS A PURE, TRANSPARENT FORM OF CALCITE, AND IS ALSO KNOWN AS ICELAND CRYSTAL.

The team then recruited 20 volunteers to take turns looking at the crystal outside on a cloudy day and measuring how accurately they could estimate the position of the hidden sun. Navigators subdivide the horizon by 360 degrees, and the team found that the volunteers could locate the sun's position to within 1 degree.

The results confirm "that the Icelandic spar is an ideal crystal, and that it can be used with great precision" for locating the sun, said ecologist Susanne Akesson of Sweden's University of Lund, who was not part of Ropars's research team.

In 2010, Akesson and her colleagues showed how local weather conditions may have influenced how light polarizes in the sky at Arctic latitudes, something Vikings would have needed to account for in their navigation. "But the question remains," she said, "whether [Icelandic spar] was in common use" in Viking times.

On that point, physics is also silent. ■

Jewelry Shows De Soto Deeper in U.S. Than Originally Thought

The path of 16th-century conquistador Hernando de Soto may need to be redrawn, thanks to a "stunning" jewelry find in Georgia.

Under a former Native American village in Georgia, deep inside what's now the United States, archaeologists say they've found 16th-century jewelry and other Spanish artifacts. The discovery suggests an expedition led by conquistador Hernando de Soto ventured far off its presumed course—which took the men from Florida to Missouri—and engaged in ceremonies with Native Americans in a thatched, pyramid-like temple.

> **TRUTH:**
> HERNANDO DE SOTO IS BEST KNOWN FOR DISCOVERING THE MISSISSIPPI RIVER.

A New Route

The discovery could redraw the map of de Soto's 1539 to 1541 march into North America, where he hoped to replicate Spain's overthrow of the Inca Empire in South America. There, the conquistador had served at the side of leader Francisco Pizarro.

A continent and five centuries away, an excavation organized by Atlanta's Fernbank Museum of Natural History found buried glass beads, iron tools, and brass and silver ornaments dating to the mid-1500s. The southern

Georgia location—where they'd been searching for a 17th-century Spanish mission—came to be called the Glass Site.

"For an Indian in the South 500 years ago, things like glass beads and iron tools might as well have been iPhones," said project leader Dennis Blanton, an independent archaeologist who until recently was Fernbank's staff archaeologist.

"These were things that were just astonishing to them. They were made of materials that were unknown and were sometimes in brilliant blue and red colors that were unmatched in the native world."

Blanton called the finding a "stunning surprise." Prior to the discovery, it had been generally accepted that de Soto and his men had crossed a river about 100 miles (160 kilometers) upstream of the site, but archaeologists hadn't suspect that the expedition had ventured so far south and east.

The trove of items—all of which could fit into a shoe box—represents the largest collection of early 16th-century Spanish artifacts ever found in the U.S. interior outside of Florida, according to Blanton, whose work was funded in part by the National Geographic Society's Committee for Research and Exploration.

Quid Pro Quo?

Excavations by Blanton's team suggest that a large building with a thatched, pyramid-shaped roof once stood at the Glass Site. The structure was

This illustration shows de Soto crossing Georgia's Ocmulgee River as Ichisi Indians watch.

surrounded by a ditch, contained a large central hearth, and may have served as an important ceremonial center or temple.

The concentration of Spanish artifacts at the Glass Site suggests de Soto may have participated in a gift-exchange ceremony with the town's chief and other leaders. It's not known what the Spaniards would have received in return, but they commonly asked for food, information, free passage, baggage carriers, and perhaps female company, Blanton said.

By comparing the archaeological results with journal accounts by the Spanish party, Blanton and his team think the Glass Site was an important village in a province ruled by the Ichisi Indians. The team also believes de Soto and his men stayed there between March 30 and April 2, 1540, according to journals.

The Man Who Fell to Earth

De Soto's party consisted of more than 600 men and hundreds of pigs and horses—animals that many of the Indians had never seen before. "There are accounts in the chronicles of how Indians at first imagined the mounted men to constitute a single creature," Blanton said.

Native American Noshes

Many of our yummiest snacks have their origins in Native American culture. Here's a small sample:

1. Chewing gum: The Aztec chewed chicle, a latex from the sapodilla tree.
2. Chocolate: Two thousand years ago the Maya cooked up Earth's first chocolate from cacao beans.
3. Vanilla: Indians in what is now Mexico first used the pods of the vanilla orchid to flavor their food.
4. Popcorn: Some Indians roasted cobs over a fire, weenie-roast style. And in South America the Moche made popcorn poppers out of pottery.

To encourage cooperation among the Indians and avoid conflict, de Soto sometimes claimed to be a god. "De Soto took advantage of the fact that the Indians revered the sun and even at Ichisi made the claim to be descended from it," Blanton said.

By 1540, rumors of an "alien people" had already spread among Native Americans in southeastern North America, but few Indians would have encountered any Europeans in the flesh, he said. "A de Soto encounter would have been for most, if not all, of the people at the Glass Site a wholly new—and undoubtedly startling—experience," Blanton said.

The fact that there is no evidence of mass killing or vandalism at the Glass Site suggests de Soto and his men were treated well during their stay, he added. And in fact Spanish journal records say the Spaniards were lavished with food and hospitality at an Ichisi village, which Blanton suspects was the Glass Site settlement.

This wasn't always the case.

"The Spaniards often treated the Natives very badly, and when the local people did not accede to their demands, de Soto would usually take the local leader hostage until he got his way," said Jeffrey Mitchem, a de Soto scholar with the Arkansas Archeological Survey, who was not involved in the discoveries. "Usually their demands for food and young women wore out their welcome very quickly," Mitchem said, "so the natives were almost always trying to make them leave as rapidly as possible."

Perhaps a Lost Colony

Mitchem agreed that the discoveries support the idea that de Soto and his men camped for several days at the Glass Site. "Many of the specific types of artifacts that have been found at [Glass Site] are the same types recovered from other sites that were contacted by the Hernando de Soto expedition," he said.

The new discoveries will not only help refine de Soto's expedition route, but could also provide valuable insight into how American Indian groups were organized in particular areas. "As we identify specific Native American towns or villages described in the narratives, we can then look at what the Spanish narratives tell us about the political situation in those specific areas," Mitchem said.

The team has also explored another Georgia Indian site, called Deer Run, but the case for a de Soto encounter there is less conclusive, Blanton said. While a visit by de Soto's party is the most likely explanation for the artifacts found at the Glass Site, Blanton says there may be another explanation: that the items were left by deserters of the lost Spanish colony of Ayllon. The settlement is known only from writings, and some scholars have proposed it was located on the Georgia coast. Though it's unlikely that the bead site harbored lost colonists, Blanton said, "if that proves to be true, then the Glass Site record is arguably even more spectacular." ■

> "For an Indian in the South 500 years ago, things like glass beads and iron tools might as well have been iPhones."
> **Dennis Blanton**
> *archaeologist*

CHAPTER 9

Natural
Pheno

mena

Ah, the wonders of nature. So peaceful. So majestic. So . . . weird? Just take a look around, and you'll see that we are surrounded by bizarre natural phenomena: giant mucuslike sea blobs in the Mediterranean Sea, enormous sinkholes that swallow three-story buildings in Guatemala, a supervolcano lurking beneath Yellowstone National Park, and a Mexican cave filled with giant crystals measuring 36 feet (11 meters) long. It looks like nature might be trying to freak us out.

Oldest Living Tree Found in Sweden

The world's oldest known living tree, a conifer that first took root at the end of the last Ice Age, has been discovered in Sweden, researchers say.

Discovered in 2004, a lone Norway spruce—of the species traditionally used to decorate European homes during Christmas—growing at an altitude of 2,985 feet (910 meters) in Dalarna Province was deemed to be the world's oldest living plant.

The roots of this tree date back 9,550 years.

How Old Is It?

The visible portion of the 13-foot-tall (4-meter-tall) "Christmas tree" isn't ancient, but its root system has been growing for 9,550 years, according to a team led by Leif Kullman, professor at Umeå University's Department of Ecology and Environmental Science in Sweden.

The tree's incredible longevity is largely due to its ability to clone itself, Kullman said. The spruce's stems or trunks have a lifespan of around 600 years, "but as soon as a stem dies, a new one emerges from the same root stock," Kullman explained. "So the tree has a very long life expectancy."

Radiocarbon Dating

Bristlecone pines in the western United States are generally recognized as the world's oldest continuously standing trees. The most ancient recorded, from California's White Mountains, is dated to around 5,000 years ago. Bristlecone pines are aged by counting tree rings, which form annually within their trunks.

But in the case of the Norway spruce, ancient remnants of its roots were radiocarbon dated. The study team also identified other ancient spruces in Sweden that were between 5,000 and 6,000 years old.

Trees much older than 9,550 years would be impossible in Sweden, because ice sheets covered the country until the end of the last Ice Age around 11,000 years ago, Kullman noted.

March of the Trees

The research forms part of an ongoing study into how and when trees colonized Scandinavia after it had thawed. "Prior to our studies the general conception was that spruce migrated to this area about 2,000 years ago, so now you will have to rewrite the textbooks," Kullman said.

Ten of the Oldest Trees in the World

1. **Methuselah, Inyo National Forest, California**—4,765 years old
2. **Zoroastrian Sarv, Abarkooh, Iran**—between 4,000 and 4,500 years old
3. **Llangernyw Yew, Llangernyw, Wales**—between 3,000 and 4,000 years old
4. **Alerce Tree, Andes Mountains, Chile**—3,620 years old
5. **The Senator, Big Tree Park, Florida**—3,500 years old
6. **General Sherman, Sequoia National Park, California**—2,300 years old
7. **Jōmon Sugi, Yakushima Island, Japan**—2,000 years old
8. **Te Matua Ngahere, Waipoua Forest, New Zealand**—2,000 years old
9. **Kongeegen, Jægerspris North Forest, Denmark**—between 1,500 and 2,000 years old
10. **Jardine Juniper, Cache National Forest, Utah**—1,500 years old

"Deglaciation seems to have occurred much earlier than generally thought," he added. "Perhaps the ice sheet during the Ice Age was much thinner than previously believed." The tree study may also help shed light on how plants will respond to current climate change, Kullman said. "We can see trees have an ability to migrate much faster than people had believed," he said.

In fact, global warming made the ancient mountain conifers easier for the study team to find. "For many millennia they survived in the mountain tundra as low-growing shrubs perhaps less than a meter high," Kullman said. "Now they are growing up like mushrooms—you can see them quite readily."

Rising Timberline

But climate change could also swamp these living Ice Age relics, he warned. The tree line has climbed up to 655 feet (200 meters) in altitude during the past century in the central Sweden study area, the team found.

"A great change in the landscape is going on," Kullman said. "Some lower mountains, which were bare tundra less than a hundred years ago, are totally covered by forest today."

Mountains tend to provide a refuge for the planet's most venerable trees because of reduced competition from neighbors and other plants and because the sparser vegetation around the timberline is less vulnerable to forest fires, Kullman said.

TRUTH: THE WORLD'S TALLEST TREE IS 379.1 FEET TALL, ABOUT AS HIGH AS 188 SCHOOL DESKS STACKED UP.

Another factor is reduced human impacts such as logging, said Tom Harlan of the Laboratory of Tree-Ring Research at the University of Arizona. "Human activity lower down has demolished all sorts of things that could have been extremely old," he said.

Harlan says the newly dated Swedish spruce trees have "quite an extraordinary age." "I have no great problems with them having a tree which has been growing there for more than 8,000 years," he said. "The date seems a little early but not out of line with other things we have seen."

For instance, Harlan noted, dead remains of Californian bristlecone pines dating to about 7,500 years ago have been found up to 500 feet (150 meters) higher in altitude than any living bristlecones. "So there was a time period then when trees were pushing aggressively into areas they had not been in before," he said.

Other tree clones may have an even more ancient lineage than the Swedish spruces, he added. Research suggests that stands of Huon pines on the Australian island of Tasmania possibly date back more than 10,000 years. ■

The Truth About Daylight Saving Time

Every spring and fall, clock confusion reigns supreme: When exactly does daylight saving time end, and why do we do it?

For most Americans, daylight saving time ends in early November, when most states fall back an hour. Time springs forward to daylight saving time in March, when daylight saving time begins again.

Where it is observed, daylight saving time has been known to cause some problems. National surveys by Rasmussen Reports, for example, show that 83 percent of respondents knew when to move their clocks ahead in spring 2010. Twenty-seven percent, though, admitted they'd been an hour early or late at least once in their lives because they hadn't changed their clocks correctly.

It's enough to make you wonder—why do we use daylight saving time in the first place?

Whose Idea Was This ?

Ben Franklin—of "early to bed and early to rise" fame—was apparently the first person to suggest the concept of daylight savings, according to computer scientist David Prerau, author of the book *Seize the Daylight: The Curious and Contentious Story of Daylight Saving Time.*

No Thanks

The federal government doesn't require U.S. states or territories to observe daylight saving time, which is why residents of Arizona, Hawaii, Puerto Rico, the Virgin Islands, American Samoa, Guam, and the Northern Marianas Islands don't need to change their clocks.

While serving as U.S. ambassador to France in Paris, Franklin wrote of being awakened at 6 a.m. and realizing, to his surprise, that the sun would rise far earlier than he usually did. Imagine the resources that might be saved if he and others rose before noon and burned less midnight oil, Franklin, tongue half in cheek, wrote to a newspaper. "Franklin seriously realized it would be beneficial to make better use of daylight but he didn't really know how to implement it," Prerau said.

It wasn't until World War I that daylight savings were realized on a grand scale. Germany was the first state to adopt the time changes, to reduce artificial lighting and thereby save coal for the war effort. Friends and foes soon followed suit. In the United States, a federal law standardized the yearly start and end of daylight saving time in 1918—for the states that chose to observe it.

During World War II, the United States made daylight saving time mandatory for the whole country, as a way to save wartime resources. Between February 9, 1942, and September 30, 1945, the government took it a step further. During this period daylight saving time was observed year-round, essentially making it the new standard time, if only for a few years.

It's hard to remember if it's "fall back" or "spring ahead."

Since the end of World War II, though, daylight saving time has always been optional for U.S. states. But its beginning and end have shifted—and occasionally disappeared. During the 1973 to 1974 Arab oil embargo, the United States once again extended daylight saving time through the winter, resulting in a 1 percent decrease in the country's electrical load, according to federal studies cited by Prerau. Thirty years later the Energy Policy Act of 2005 was enacted, mandating a controversial month-long extension of daylight saving time, starting in 2007.

But does daylight saving time really save any energy?

Energy Saver or Waster?

In recent years several studies have suggested that daylight saving time doesn't actually save energy—and might even result in a net loss. Environmental economist Hendrik Wolff, of the University of Washington, co-authored a paper that studied Australian power-use data when parts of the country extended daylight saving time for the 2000 Sydney Olympics and others did not. The researchers found that the practice reduced lighting and electricity consumption in the evening but increased energy use in the now dark mornings—wiping out the evening gains.

Likewise, Matthew Kotchen, an economist at the University of California, saw in Indiana a situation ripe for study. Prior to 2006, only 15 of the state's 92 counties observed daylight saving time. So when the whole state adopted daylight saving time, it became possible to compare before-and-after energy use. While use of artificial lights dropped, increased air-conditioning use more than offset any energy gains, according to the daylight saving time research Kotchen led for the National Bureau of Economic Research in 2008.

That's because the extra hour that daylight saving time adds in the evening is a hotter hour. "So if people get home an hour earlier in a warmer house,

> "Light doesn't do the same things to the body in the morning and the evening. More light in the morning would advance the body clock, and that would be good. But more light in the evening would even further delay the body clock."
> **Till Roenneberg**
> *chronobiologist,*
> *Ludwig-Maximillians University,*
> *Munich, Germany*

they turn on their air conditioning," the University of Washington's Wolff said. In fact, Hoosier consumers paid more on their electric bills than before they made the annual switch to daylight saving time, the study found.

But other studies do show energy gains. In an October 2008 daylight saving time report to Congress, mandated by the same 2005 energy act that extended daylight saving time, the U.S. Department of Energy asserted that springing forward does save energy. Extended daylight saving time saved 1.3 terawatt hours of electricity. That figure suggests that daylight saving time reduces annual U.S. electricity consumption by 0.03 percent and overall energy consumption by 0.02 percent. While those percentages seem small, they could represent significant savings because of the nation's enormous total energy use.

Winners and Losers

What's more, savings in some regions are apparently greater than in others. California, for instance, appears to benefit most from daylight saving time—perhaps because its relatively mild weather encourages people to stay outdoors later. The Energy Department report found that daylight saving time resulted in an energy savings of 1 percent daily in the state.

But Wolff, one of many scholars who contributed to the federal report, suggested that the numbers were subject to statistical variability and shouldn't be taken as hard facts.

And daylight savings' energy gains in the United States largely depend on your location in relation to the Mason-Dixon Line, Wolff said. "The North might be a slight winner, because the North doesn't have as much air conditioning," he said. "But the South is a definite loser in terms of energy consumption. The South has more energy consumption under daylight saving."

Healthy or Harmful

For decades, advocates of daylight savings have argued that, energy savings or no, daylight saving time boosts health by encouraging active lifestyles—a claim Wolff and colleagues are currently putting to the test.

"In a nationwide American time-use study, we're clearly seeing that, at the time of daylight saving time extension in the spring, television watching is substantially reduced and outdoor behaviors like jogging, walking, or going to the park are substantially increased," Wolff said.

But others warn of ill effects. Till Roenneberg, a chronobiologist at Ludwig-Maximilians University in Munich, Germany, said his studies show that our bodies never adjust to gaining an "extra" hour of sunlight.

"The consequence of that is that the majority of the population has drastically decreased productivity, decreased quality of life, increasing susceptibility to illness, and is just plain tired," Roenneberg said. One reason so many people in the developed world are chronically overtired, he said, is that their optimal circadian sleep periods are out of whack with their actual sleep schedules.

Shifting daylight from morning to evening only increases this lag, he said. "Light doesn't do the same things to the body in the morning and the evening. More light in the morning would advance the body clock, and that would be good. But more light in the evening would even further delay the body clock."

> # TRUTH:
> ## EVERY DAY IS ABOUT 55 BILLIONTHS OF A SECOND LONGER THAN THE DAY BEFORE IT.

Pros and Cons

With verdicts on the benefits, or costs, of daylight savings so split, it may be no surprise that the yearly time changes inspire polarized reactions. In the United Kingdom, for instance, the Lighter Later movement—part of 10:10, a group advocating cutting carbon emissions—argues for a sort of extreme daylight savings. First, they say, move standard time forward an hour, then keep observing daylight saving time as usual—adding two hours of evening daylight to what we currently consider standard time.

The folks behind standardtime.com, on the other hand, want to abolish daylight saving time altogether. Calling energy-efficiency claims "unproven," they write: "If we are saving energy let's go year round with Daylight Saving Time. If we are not saving energy let's drop Daylight Saving Time!"

But don't most people enjoy that extra evening sun every summer? Even that remains in doubt. National telephone surveys by Rasmussen Reports from fall 2009 and spring 2010 deliver the same answer: Most people just "don't think the time change is worth the hassle." Forty-seven percent agreed with that statement, while only 40 percent disagreed.

But *Seize the Daylight* author David Prerau said his research on daylight saving time suggests most people are fond of it. "I think the first day of daylight saving time is really like the first day of spring for a lot of people," Prerau said. "It's the first time that they have some time after work to make use of the springtime weather. I think if you ask most people if they enjoy having an extra hour of daylight in the evening eight months a year, the response would be pretty positive." ∎

UFO-Like Clouds

Linked to Military Maneuvers?

Three weird clouds appeared in the sky over South Carolina in January 2011, and a local photographer snapped a picture. After he posted it on the Internet, it seems everyone had a theory about them. But what's the real story?

Three nearly identical, UFO-like cloud formations appeared over Myrtle Beach, South Carolina, in January 2011, sparking online discussions linking the features to everything from the Second Coming to recent mass bird deaths to secret military experiments. At least one scientist believes the so-called hole-punch clouds have a military explanation, though it may not be quite what conspiracy theorists expect.

What Are Little Clouds Made Of?

Earth isn't the only planet covered with clouds. Earth's sister planet Venus has them, too. Instead of being composed of water droplets and ice crystals, though, Venus's clouds are made of sulfuric acid, and would be deadly to humans.

The Clouds Appear

On January 7, 2011, IT technician Wesley Tyler was running out to his car for a computer part when he noticed the saucerlike formations. "At first we thought they were tornado clouds, but the air was so still—like mausoleum still," Tyler said. "You just knew it was unusual. I've lived on the beach for years and never seen anything like that."

Back home, he uploaded pictures of the clouds to Facebook, tagging a meteorologist friend, who later identified the phenomena as hole-punch clouds, or punch-hole clouds.

Hole-Punch Clouds

Hole-punch clouds are miniature snowstorms that can occur in thin, sub-freezing cloud layers. The lack of fine particles, such as dust, in the clouds means water droplets have little to condense around, so they don't turn to ice until the cloud hits about minus 38 degrees Fahrenheit (minus 36 degrees Celsius).

"Basically, the water molecules become sluggish enough at this temperature to form their own cluster of ice that produces an ice crystal spontaneously," according to ice microphysicist Andrew Heymsfield.

When airplanes ascend into this type of cloud, the rearward force created by propellers or by air forced over wings causes air to expand. This expansion can cool a vaguely circular section of the cloud to the point where many of the water droplets freeze and ice crystals form, according to a June hole-punch cloud study co-authored by Heymsfield in the *Bulletin of the American Meteorological Society*.

Over the next 45 minutes or so, ice crystals grow and spread outward, often resulting in a tightly contained, roughly half-hour snowstorm—leaving behind a hole "punched" in the cloud.

A hole-punch cloud over Mobile, Alabama

Conspiracy Theory or Divine Message?

Tyler, the photographer, was skeptical of the airplane explanation, due to the sheer number and close proximity of the cloud formations. "I've scoured the Internet and have yet to find more than one hole-punch cloud in a single frame," he said.

Myrtle Beach International, he added, is "not that busy an airport." And, he said, "I've read that these clouds form at 20,000 feet [6,100 meters], and these clouds looked like they were right above us. "I doubt they were created by airplanes," Tyler concluded—and he's not alone.

After his pictures were posted on SpaceWeather.com, the Myrtle Beach resident began hearing from people all over the world. Some suspected a more colorful cause—perhaps the military-funded High Frequency Active Auroral Research Program, or HAARP, which conspiracy theorists have linked to earthquakes, chronic fatigue syndrome, global warming, and other phenomena.

Though remote, the observatory-and-antenna facility in Gakona, Alaska, is anything but secret. Even so, its use of radio waves to "excite" areas of Earth's ionosphere has helped convince some that HAARP can control weather—and perhaps even create triple hole-punch clouds.

> "The hole sizes and the structure of the snow falling out of the holes suggest that all three holes were made at nearly the same time. My suspicion is that military aircraft were flying in formation or one behind the other."
>
> **Andrew Heymsfield**
> *ice microphysicist, on the more likely cause of the odd cloud formations*

"There is no doubt," one HAARP theorist wrote of the Myrtle Beach apparition on the Big Wobble message board, "it's an electromagnetic corridor produced by our technology." Another wrote on starseeds.net, "This could be related to HAARP or some weather manipulation as it also ties in with the bird deaths."

And on Rapture in the Air, a site devoted to signs of the Second Coming of Christ, "mike" wrote, "Hope the photos was taken after 3 invisible space [arks] came down from heaven which the Lord has sent to earth . . ."

While Tyler doesn't necessarily buy these theories, he thought the airplane explanation was flawed. "There must be another explanation—natural or otherwise."

Military Explanation

To Heymsfield, the physicist, the explanation is both natural and otherwise. "To me, it's a slam dunk" that these are hole-punch clouds that were created the usual way—by planes—said Heymsfield, of the National Center for Atmospheric Research in Boulder, Colorado.

There's "nothing at all" surprising about the picture, he added. For one thing, it's the right type of cloud—thin, with no other layers above it—as evidenced by the clear skies just beyond, he said. And the cloud layer's temperature fits the hole-punch model: 14 degrees Fahrenheit (minus 10 degrees Celsius), according to National Weather Service records.

As for the cloud being low in the sky—9,000 feet (2,700 meters), according to the weather service—"it doesn't matter," as long as the cloud layer is cold enough, he said.

But why three together? "The hole sizes and the structure of the snow falling out of the holes suggest that all three holes were made at nearly the same time," he said. "My suspicion is that military aircraft were flying in formation or one behind the other."

Training Planes

And in fact, it's "very common" for training maneuvers to take place over Myrtle Beach, according to Robert Sexton, community relations manager for nearby Shaw Air Force Base. More to the point, Sexton confirmed that fighter jets from Shaw and from the South Carolina Air National Guard's 169th Fighter Wing were training off the South Carolina coast on January 7 between 9 a.m. and 2 p.m.

"After us, the Marines were in the airspace from 3 to 4 p.m. with F-18s" out of the Marine Corps air station in Beaufort, South Carolina, he emailed.

After having heard the new evidence, Tyler, the photographer, said he's convinced by the aircraft explanation, though he initially seemed slightly disappointed by its straightforwardness. But "that's still cool enough," Tyler decided. "I'm a conspiracist, but also a naturalist." ■

> **TRUTH:**
> HOLE-PUNCH CLOUDS ARE FORMED WHEN AIRPLANES PASS THROUGH CLOUDS IN THE RIGHT ATMOSPHERIC CONDITIONS AND MAKE LIQUID WATER DROPLETS FREEZE AND DROP AS SNOW, LEAVING A CIRCULAR FISSURE.

Giant Sinkholes

Pierce Guatemala

In 2007 and again in 2010, giant sinkholes opened up beneath the streets of Guatemala City. What could have caused these disasters?

A huge sinkhole in Guatemala City, Guatemala, crashed into being in 2010, reportedly swallowing a three-story building—and echoing a similar, 2007 occurrence.

Explaining Sinkholes

Sinkholes are natural depressions that can form when water-saturated soil and other particles become too heavy and cause the roofs of existing voids

The 2010 sinkhole in Guatemala had likely been forming for several weeks or even years before floodwaters caused it to cave in.

in the soil to collapse. Another way sinkholes can form is if water enlarges a natural fracture in a limestone bedrock layer. As the crack gets bigger, the topsoil gently slumps, eventually leaving behind a sinkhole.

Sinkholes are particularly prevalent when heavy rains follow a long period of drought, said Jonathan Martin, a geologist at the University of Florida. Drought can empty subterranean cavities of water, making them less able to support the overlying soil—flooding only adds to the danger. "If there's a tropical storm and all of a sudden the soil above the cavity is filled with water instead of air, the weight will cause the [sinkhole] to collapse," Martin said.

Although the exact mechanism behind it is unclear, the 2010 sinkhole had likely been weeks or even years in the making—floodwaters from tropical storm Agatha caused it to finally collapse, scientists say. The sinkhole appears to be about 60 feet (18 meters) wide and about 30 stories deep, said James Currens, a hydrogeologist at the University of Kentucky.

> # TRUTH:
> ## IN THE U.S., THE MOST DAMAGE FROM SINKHOLES TENDS TO OCCUR IN FLORIDA, TEXAS, ALABAMA, MISSOURI, KENTUCKY, TENNESSEE, AND PENNSYLVANIA.

Sewer Problems

A ruptured sewer line is thought to have caused the sinkhole that appeared in Guatemala City in 2007. The 2010 Guatemala sinkhole could have formed in a similar fashion, Currens said. A burst sanitary or storm sewer may have been slowly saturating the surrounding soil for a long time before tropical storm Agatha added to the inundation. "The tropical storm came along and would have dumped even more water in there, and that could have been the final trigger that precipitated the collapse," Currens said.

Depending on the makeup of the subsurface layer, the sinkhole "could eventually enlarge and take in more buildings," he said. Typically, officials fill in sinkholes with large rocks and other debris. But this one "is so huge that it's going to take a lot of fill material to fill it," Currens said. "I don't know what they're going to do." ■

Yellowstone Has Bulged

as Magma Pocket Swells

A supervolcano beneath Yellowstone National Park has been taking some deep "breaths" lately, causing the ground to rise as much as 10 inches (25 cm). Is a super-eruption far behind?

Yellowstone National Park's supervolcano took a deep "breath" in 2011, causing miles of ground to rise dramatically, scientists report.

A Sleeping Giant

The simmering volcano has produced major eruptions—each a thousand times more powerful than Mount St. Helens's 1980 eruption—three times in the past 2.1 million years. Yellowstone's caldera, which covers a 25- by 37-mile (40- by 60-kilometer) swath of Wyoming, is an ancient crater formed after the last big blast, some 640,000 years ago. Since then, about 30 smaller eruptions—including one as recent as 70,000 years ago—have filled the caldera with lava and ash, producing the relatively flat landscape we see today.

> **TRUTH:**
> BOLTS OF LIGHTNING CAN SHOOT OUT OF AN ERUPTING VOLCANO.

But beginning in 2004, scientists saw the ground above the caldera rise upward at rates as high as 2.8 inches (7 centimeters) a year. The rate slowed between 2007 and 2010 to a centimeter a year or less. Still, since the start of the swelling, ground levels over the volcano have been raised by as much as 10 inches (25 centimeters) in places. "It's an extraordinary uplift, because

Castle Geyser in Yellowstone National Park

it covers such a large area and the rates are so high," said the University of Utah's Bob Smith, a longtime expert in Yellowstone's volcanism.

Predicting the Next "Burp"

Scientists think a swelling magma reservoir 4 to 6 miles (7 to 10 kilometers) below the surface is driving the uplift. Fortunately, the surge doesn't seem to herald an imminent catastrophe, Smith said. "At the beginning, we were concerned it could be leading up to an eruption," said Smith, who co-authored a paper on the surge published in the December 3, 2010, edition of *Geophysical Research Letters*.

"But once we saw [the magma] was at a depth of 10 kilometers, we weren't so concerned. If it had been at depths of 2 or 3 kilometers [1 or 2 miles], we'd have been a lot more concerned." Studies of the surge, he added, may offer valuable clues about what's going on in the volcano's subterranean plumbing, which may eventually help scientists predict when Yellowstone's next volcanic "burp" will break out.

Taking Regular Breaths

Smith and colleagues at the U.S. Geological Survey (USGS) Yellowstone Volcano Observatory have been mapping the caldera's rise and fall using tools such as global positioning systems (GPS) and interferometric synthetic aperture radar (InSAR), which gives ground-deformation measurements.

Ground deformation can suggest that magma is moving toward the surface before an eruption: The flanks of Mount St. Helens, for example, swelled dramatically in the months before its 1980 explosion. But there are also many examples, including the Yellowstone supervolcano, where it appears the ground has risen and fallen for thousands of years without an eruption.

Liquid Hot Magma

According to current theory, Yellowstone's magma reservoir is fed by a plume of hot rock surging upward from Earth's mantle. When the amount of magma flowing into the chamber increases, the reservoir swells like a lung and the surface above expands upward. Models suggest that during the recent uplift, the reservoir was filling with 0.02 cubic miles (0.1 cubic kilometer) of magma a year. When the rate of increase slows, the theory goes, the magma likely moves off horizontally to solidify and cool, allowing the surface to settle back down.

Based on geologic evidence, Yellowstone has probably seen a continuous cycle of inflation and deflation during the past 15,000 years, and the cycle will likely continue, Smith said. Surveys show, for example, that the caldera rose some 7 inches (18 centimeters) between 1976 and 1984 before dropping back about 5.5 inches (14 centimeters) during the next decade.

"These calderas tend to go up and down, up and down," he said. "But every once in a while they burp, creating hydrothermal explosions, earthquakes, or—ultimately—they can produce volcanic eruptions."

Volcanoes and Geysers and Quakes, Oh My!

Predicting when an eruption might occur is extremely difficult, in part because the fine details of what's going on under Yellowstone are still undetermined. What's more, continuous records of Yellowstone's activity have been made only since the 1970s—a tiny slice of geologic time—making it hard to draw conclusions. "Clearly some deep source of magma feeds Yellowstone, and since Yellowstone has erupted in the recent geological past, we know that there is magma at shallower depths too," said Dan Dzurisin, a Yellowstone expert with the USGS Cascades Volcano Observatory in Washington State.

"There has to be magma in the crust, or we wouldn't have all the hydrothermal activity that we have," Dzurisin added. "There is so much heat coming out of Yellowstone right now that if it wasn't being reheated by magma, the whole system would have gone stone cold since the time of the last eruption 70,000 years ago."

The large hydrothermal system just below Yellowstone's surface, which produces many of the park's top tourist attractions, may also play a role in ground swelling, Dzurisin said, though no one is sure to what extent. "Could it be that some uplift is caused not by new magma coming in but by the hydrothermal system sealing itself up and pressurizing?" he asked. "And then it subsides when it springs a leak and depressurizes? These details are difficult."

Feel It Shakin'

And it's not a matter of simply watching the ground rise and fall. Different areas may move in different directions and be interconnected in unknown ways, reflecting the as yet unmapped network of volcanic and hydrothermal plumbing. The roughly 3,000 earthquakes in Yellowstone each year may offer even more clues about the relationship between ground uplift and the magma chamber. For example, between December 26, 2008, and January 8, 2009, some 900 earthquakes occurred in the area around Yellowstone Lake.

This earthquake "swarm" may have helped to release pressure on the magma reservoir by allowing fluids to escape, and this may have slowed the rate of uplift, the University of Utah's Smith said. "Big quakes [can have] a relationship to uplift and deformations caused by the intrusion of magma," he said. "How those intrusions stress the adjacent faults, or how the faults might transmit stress to the magma system, is a really important new area of study."

Overall, USGS's Dzurisin added, "The story of Yellowstone deformation has gotten more complex as we've had better and better technologies to study it." ∎

Supervolcanoes of the World

1. **Lake Toba, Sumatra, Indonesia—** The 1,080-square-mile Toba caldera can be described as Yellowstone's "big sister." It erupted 74,000 years ago, triggering a global cold spell.
2. **Long Valley, California—**At 200 square miles, the Long Valley caldera's most recent eruption was in Mono Lake, 250 years ago.
3. **Lake Taupo, New Zealand—**A massive eruption created Lake Taupo 26,500 years ago, when the caldera collapsed and filled up with water. The 485-square-mile caldera erupted again in the year A.D. 181.
4. **Valles Caldera, New Mexico—**The 175-square-mile Valles caldera last erupted 1.2 million years ago. Hot springs around it remain active.
5. **Aira, Japan—**On January 10, 1914, the Sakura-jima volcano, which forms part of the 150-square-mile Aira caldera, became active and caused hundreds of earthquakes. Two days later, it erupted with ash, steam, and lava.

Brand-New Cloud

Choppy, undulating clouds could be examples of the first new type of cloud to be recognized since 1951. Or so hopes Gavin Pretor-Pinney, founder of the Cloud Appreciation Society.

In 2005 British cloud enthusiast Gavin Pretor-Pinney said he began getting photos of "dramatic" and "weird" clouds that he didn't know how to define. In 2009, he began preparing to propose the odd formations as a new cloud variety to the UN's World Meteorological Organization, which classifies cloud types.

Enter *Undulatus Asperatus*

Pretor-Pinney jokingly calls it the "Jacques Cousteau cloud," after its resemblance to a roiling ocean surface seen from below. But the cloud fan has proposed a "formal," Latin name: *Undulatus asperatus*—roughly, "a very turbulent, violent, chaotic form of undulation," explained Pretor-Pinney, author of the new *Cloud Collector's Handbook.*

> **TRUTH:**
> **A CLOUD CAN WEIGH MORE THAN A MILLION POUNDS.**

Margaret LeMone, a cloud expert with the National Center for Atmospheric Research in Boulder, Colorado, said that she has taken photos of asperatus clouds intermittently throughout the past 30 years. It's likely that the cloud will turn out to be a new variety, LeMone said. "Having a group of people enthusiastic about clouds can only help the field of meteorology," she added.

Asked how such a striking cloud type could go unrecognized, Pretor-Pinney cites its rarity—and the proliferation and portability of digital cameras. "Technology has allowed us to have this new perspective on the sky."

Wild Cards

This apparently new class of cloud is still a mystery. But experts suspect asperatus clouds' choppy undersides may be due to strong winds disturbing previously stable layers of warm and cold air. Asperatus clouds may spur the first new classification in the World Meteorological Organization's *International Cloud Atlas* since the 1950s, Pretor-Pinney said.

Since the last addition to the atlas, the emergence of satellite imagery has pushed meteorologists to take a much broader view on weather and focus less on small-scale cloud formations. But "the tide is turning back again," in part because the humble cloud is seen as a "wild card" in climate-change prediction, Pretor-Pinney said.

LeMone agreed that clouds are a "big unknown" in climate change, mostly because climate-change models do not provide a high-enough resolution to determine what clouds' impacts will be on a changing world.

> "Even if you live in the middle of the city, the sky is the last wilderness you can look out on."
> **Gavin Pretor-Pinney**
> *cloud enthusiast and author of* Cloud Collector's Handbook

A Bad Rap

Gavin Pretor-Pinney, who is proposing that asperatus clouds be officially recognized, said that clouds get a "bad rap." "People complain about . . . having a cloud hanging over them, compared to someone having a sunny outlook on life," said Pretor-Pinney. "To me, clouds are one of the most beautiful parts of nature." ■

Solved! The Case of the Giant Crystals

A team of geologists have solved the mystery behind the formation of giant crystals in a Mexican cave known as the "Sistine Chapel of crystals."

TRUTH: INSIDE THE CAVE OF CRYSTALS, THE AVERAGE TEMPERATURE IS 122 DEGREES FAHRENHEIT, WITH ALMOST 100 PERCENT HUMIDITY.

Buried a thousand feet (300 meters) below Naica Mountain in the Chihuahuan Desert, the spectacular wonder was discovered by two miners excavating a new tunnel for the Industrias Peñoles company in 2000. They found a cave filled with some of the largest natural crystals ever found: translucent gypsum beams measuring up to 36 feet (11 meters) long and weighing up to 55 tons.

"It's a natural marvel," said Juan Manuel García-Ruiz, of the University of Granada in Spain.

It took seven years for geologist García-Ruiz and a team of researchers to unlock the mystery of just how the minerals in Mexico's Cueva de los Cristales (Cave of Crystals) achieved their monumental forms.

Massive beams of selenite dwarf an explorer in the Cave of Crystals.

Underwater Wonder

To learn how the crystals grew to such gigantic sizes, García-Ruiz studied tiny pockets of fluid trapped inside. The crystals, he said, thrived because they were submerged in mineral-rich water with a very narrow, stable temperature range—around 136 degrees Fahrenheit (58 degrees Celsius). At this temperature the mineral anhydrite, which was abundant in the water, dissolved into gypsum, a soft mineral that can take the form of the crystals in the Naica cave.

Caves of Swords and Crystals

The mining complex in Naica contains some of the world's largest deposits of silver, zinc, and lead. In 1910, miners discovered another spectacular cavern beneath Naica.

Its walls studded with crystal "daggers," the Cave of Swords is closer to the surface, at a depth of nearly 400 feet (120 meters). While there are more crystals in the upper cave, they are far smaller, typically about a yard (a meter) long.

The Cave of Crystals is a horseshoe-shaped cavity in limestone rock about 30 feet (10 meters) wide and 90 feet (30 meters) long. Its floor is covered in crystalline, perfectly faceted blocks. The huge crystal beams jut out from both the blocks and the floor. "There is no other place on the planet where the mineral world reveals itself in such beauty," García-Ruiz said.

Looters, Beware

The Cave of Crystals' stifling temperatures and the fact that it takes 20 minutes to drive to its entrance through a twisting mine shaft hasn't prevented looters from trying to get a piece of the treasure. One of the crystals was found with a deep scar where someone tried, but failed, to cut through it. Subsequently, the cave has been supplied with a heavy steel door.

Cooling Off Creates Crystals

Volcanic activity that began about 26 million years ago created Naica Mountain and filled it with high-temperature anhydrite, which is the anhydrous—lacking water—form of gypsum. Anhydrite is stable above 136 degrees Fahrenheit (58 degrees Celsius). Below that temperature gypsum is the stable form.

When magma underneath the mountain cooled and the temperature dropped below 58 degrees Celsius, the anhydrite began to dissolve. The anhydrite slowly enriched the waters with sulfate and calcium molecules, which for millions of years have been deposited in the caves in the form of huge selenite gypsum crystals. "There is no limit to the size a crystal can reach," García-Ruiz said.

But, he said, for the Cave of Crystals to have grown such gigantic crystals, it must have been kept just below the anhydrite-gypsum transition temperature for many hundreds of thousands of years. In the upper cave, by contrast, this transition temperature may have fallen much more rapidly, leading to the formation of smaller crystals.

To Reflood or Not to Reflood

While the chance of this set of conditions occurring in other places in the world is remote, García-Ruiz expects that there are other caves and caverns at Naica containing similarly large crystals. "The caves containing larger crystals will be located in deeper levels with temperatures closer to, but no higher than, 58 degrees Celsius," he said.

> **TRUTH:**
> THE WORLD'S LARGEST KNOWN CRYSTAL IS 37.4 FEET LONG. THAT'S EIGHT TIMES TALLER THAN THE AVERAGE TEN-YEAR-OLD.

He has recommended to the mining company that the caves should be preserved.

The only reason humans can get into the caves today, however, is because the mining company's pumping operations keep them clear of water. If the pumping is stopped, the caves will again be submerged and the crystals will start growing again, García-Ruiz said.

So what happens if—or when—the mine is closed?

"That's an interesting question," García-Ruiz said. "Should we continue to pump water to keep the cave available so future generations may admire the crystals? Or should we stop pumping and return the scenario to the natural origin, allowing the crystals to regrow?" ■

Giant, Mucuslike Sea Blobs

On the Rise, Pose Danger

As sea temperatures have risen in recent decades, enormous sheets of a mucuslike material have begun forming more often, oozing into new regions, and lasting longer, a new Mediterranean Sea study says. And the blobs may be more than just unpleasant.

Beware of the blob—this time, it's for real. Up to 124 miles (200 kilometers) long, the mucilages appear naturally, usually near Mediterranean coasts in the summer. The season's warm weather makes seawater more stable, which facilitates the bonding of the organic matter that makes up the blobs. Now, due to warmer temperatures, the mucilages are forming in winter too—and lasting for months.

> **"The suit was impossible to wash totally, because it was covered by a layer of greenish slime."**
> **Serena Ford Umani**
> *co-author of mucilage study, on the state of her wet suit after diving into the blob*

More than a Nuisance

Until now, the light-brown "mucus" was seen as mostly a nuisance, clogging fishing nets and covering swimmers with a sticky gel—newspapers from the 1800s show beachgoers holding their noses, according to study leader Roberto

Danovaro, director of the Marine Science Department at the Polytechnic University of Marche in Italy.

But a study found that Mediterranean mucilages harbor bacteria and viruses, including potentially deadly *E. coli,* Danovaro said. Those pathogens threaten human swimmers as well as fish and other sea creatures, according to the report, published in the journal *PloS One.*

Building the Perfect Blob

A mucilage begins as "marine snow": clusters of mostly microscopic dead and living organic matter, including some life-forms visible to the naked eye—small crustaceans such as shrimp and copepods, for example. Over time, the snow picks up other tiny hitchhikers, looking for a meal or safety in numbers, and may grow into a mucilage.

The blobs were first identified in 1729 in the Mediterranean, where they're most often seen. The sea's relative stillness and shallowness make the water column more stable, providing ideal conditions for mucilage formation. For this study, Danovaro and colleagues studied historical reports of mucilage in the Mediterranean from 1950 to 2008. Outbreaks, they discovered, were more likely when sea-surface temperatures were warmer than average.

TRUTH: MUCILAGES WERE FIRST IDENTIFIED IN THE MEDITERRANEAN IN 1729.

Smelly Green Slime

In 1991, Italian marine biologist Serena Fonda Umani swam alongside a mucilage—the mass is too dense to swim inside—in the Adriatic Sea, an arm of the Mediterranean. She remembers diving about 50 feet (15 meters) down when she got the sensation of a ghost floating over her—"sort of an alien experience."

Umani, a co-author of the new study with Danovaro and Antonio Pusceddu, of the Polytechnic University of Marche, has also dived into marine snow—the mucilage's precursor. She described it as being like swimming through a sugar solution. Out of the water, the dried "sugar" stiffened her hair and stuck to her wet suit. "The suit was impossible to wash totally, because it was covered by a layer of greenish slime," said Umani, of Italy's University of Trieste. "It was a nightmare."

Few people would purposely swim into a mucilage, said Farooq Azam, a marine microbiologist at the University of California's Scripps Institution of Oceanography. "If you were not familiar with this—and especially if you were familiar—you wouldn't want to go near it," said Azam, who was not involved in the new study. A giant odiferous blob drifting offshore is "certainly not the seascape that one goes to the beach [for]," Azam added.

Health Hazard

Umani and colleagues sampled coastal waters and mucilage from the Adriatic in 2007. The study team discovered that the blobs are hot spots for viruses and bacteria, including the deadly *E. coli*. Study leader Danovaro said, "Now we see that . . . the release of pathogens from the mucilage can be potentially problematic" for human health.

Fish and other marine animals that have no choice but to swim with mucilages are most vulnerable to their disease-carrying bacteria, the study says. The noxious masses can also trap animals, coating their gills and suffocating them, Danovaro said. And the biggest blobs can sink to the bottom, acting like a huge blanket that smothers life on the seafloor.

Mucilages aren't a concern for just the Mediterranean, Danovaro added. Recent studies tentatively suggest that mucus may be spreading throughout oceans from the North Sea to Australia, perhaps because of rising temperatures, he said.

"It's a good example [of what will happen if] we don't do something to stop climate warming," Danovaro said. "There are consequences [if] we continue to deny the scientific evidence."

Beyond warm temperatures, it's still not exactly clear what drives the blobs' formation, Scripps' Azam pointed out. For instance, no one knows why the dead marine matter in the blobs doesn't decompose. "It's important we do find out" what's driving the rise of the blobs, Azam said, "for the sake of the rest of the worlds' oceans." ∎

Beach Bacteria

In 2005, the Clean Beaches Council, a Washington, D.C.-based advocacy group, issued a report that the sand at many U.S. beaches contains bacteria that indicate potentially unhealthy levels of fecal material. The indicator bacteria, which include a benign form of *E. coli,* pose little health risk to people, but still serve as warning signs that harmful fecal microorganisms may be present as well. The report is meant to raise awareness for beachgoers to "leave no trace" when visiting the beach and to wash up after playing in the sand.

"Bodies" Fill Underwater Sculpture Park

Off the shores of Cancún, Mexico, stand hundreds of statues, a new underwater sculpture garden that doubles as an artificial reef.

More than 400 of the permanent sculptures have been installed in 2010 in the National Marine Park of Cancún, Isla Mujeres and Punta Nizuc as part of a major artwork called "The Silent Evolution." The installation is the first endeavor of a new underwater museum called MUSA, or Museo Subacuático de Arte.

Created by Mexico-based British sculptor Jason deCaires Taylor, the Caribbean installation is intended to eventually cover more than 4,520 square feet (420 square meters), which would make it "one of the largest and most ambitious underwater attractions in the world," according to a museum statement. Along with creator Taylor, a team of artists, builders, marine biologists, engineers, and scuba divers are working together to complete it.

> **TRUTH:**
> **THE SCULPTURES ARE MADE FROM CEMENT, SAND, MICROSILICA, FIBERGLASS, AND LIVE CORAL.**

In doing so, Taylor hopes the reefs, which are already stressed by marine pollution, warming waters, and overfishing, can catch a break from the approximately 750,000 tourists who visit local reefs each year. "That puts a

THE HUMAN REEF

1: *A snorkeler swims* over life-size statues near Cancún, Mexico, in a late 2010 picture.

2: *Before being taken underwater,* the sculptures stand on a Cancún, Mexico, beach in September 2010.

3: Pictured in late 2010, *"Sarah,"* modeled after a U.K. linguistics professor, is the only statue with a false lung.

4: Jason deCaires Taylor works on a cast of *Charlie Brown,* a 67-year-old Mexican fisherman. Brown "was the only person to fall asleep during the casting process," Taylor reported.

5: The people in *"The Silent Evolution"* were created from live casts of a wide sample of people, most of them locals—including Lucky, a Mexican carpenter (center), according to Taylor.

1

2

lot of pressure on the existing reefs," Taylor told National Geographic News. "So part of this project is to actually discharge those people away from the natural reefs and bring them to an area of artificial reefs."

School of Rock

The sculptures are made of a special kind of marine cement that attracts the growth of corals, according to creator Taylor. That in turn encourages fish and other marine life to colonize the reef, he said.

"Already, I think there're a thousand different fish living on them. There're lobsters, there're big schools of angelfish. And there's a big coating of algae, which is one of the [first] things to settle."

A Way to Recycle

Along the Atlantic coast from New Jersey to Georgia, subway cars—along with armored tanks, naval ships, tugboats, and a large number of concrete culverts—have been strategically dumped in the ocean to act as artificial reefs. Most of the mid-Atlantic ocean surface is featureless sand punctuated with mud splotches, so the artificial reefs help boost marine life and, by extension, keep recreational fishers happy.

Cast of Characters

The people in "The Silent Evolution" were created from live casts of a wide sample of people, most of them locals—including Lucky, a Mexican carpenter; a 3-year-old boy, Santiago; and an 85-year-old nun, Rosario. Also depicted are an accountant, yoga instructor, and acrobat, among others.

"Sarah," modeled after a U.K. linguistics professor, is the only "Silent Evolution" statue with a false lung, according to Taylor. Divers can either fill the lung by blowing bubbles into a hole on her back or using air from their tanks. The air then slowly escapes though the opening in her mouth.

The tight gathering of people is meant to illustrate "how we are all facing serious questions concerning our environment and our impact on the natural world," according to a museum statement.

Bottom Dwellers

The sculptures were lowered into the waters off Cancún in late 2010. There, they sit in just 30 feet (9 meters) of water, which allows visitors in glass-bottomed boats to observe the artwork, according to a museum statement.

Boatbound visitors can also see big schools of fish above the statues, Taylor said. "If there're any sorts of predators or any danger, [the fish] sort of drop below and then hide out in the [statues'] feet area."

The builders of "The Silent Evolution" hope to usher in a new age of responsible tourism in the area, according to the museum. MUSA, the underwater museum, plans to add sculptures as funding becomes available. But "The Silent Evolution" won't ever really be finished, since marine life will continue adding its own touches for centuries.

The cement figures will change in appearance in time as coral and other marine life takes over—all part of Taylor's vision. "The manifestation of living organisms cohabiting and ingrained in our being is intended to remind us of our close dependency on nature and the respect we should afford it," according to a museum statement.

> **TRUTH:** UPON THE INSTALLATION'S COMPLETION, THE WEIGHT OF THE STATUES WILL TOTAL MORE THAN 180 TONS.

Balancing Act

Already the exhibition is drawing more divers, and area dive-tour providers are hoping the underwater museum boosts business and supports reef health, according to a museum statement. "This is a perfect balance where we are protecting the reef, where we are bringing the tourists into the natural area," Roberto Diaz, president of both the Cancún Nautical Association and the museum, told National Geographic News.

"We are providing art to make it beautiful, and altogether [it] will help." ∎

Moss Has Cloned Itself

for 50,000 Years, Study Says

A moss spreading throughout the Hawaiian Islands appears to be an ancient clone that has copied itself for some 50,000 years—and may be one of the oldest multicellular organisms on Earth, a new study suggests.

Once only found on Hawaii's Big Island, an ancient moss has begun to appear throughout the Hawaiian Islands. Studies show that this little green organism has been cloning itself for around 50,000 years.

I Think I'm a Clone Now

The peat moss *Sphagnum palustre* is found throughout the Northern Hemisphere, but the moss living in Hawaii appears to reproduce only through cloning. All the moss populations sampled share a rare genetic marker, which suggests they're descended from a single founder plant that was carried via wind to Hawaii tens of thousands of years ago.

Surprisingly Diverse

A genetic analysis of the *S. palustre* moss in Hawaii reveals surprising diversity, which challenges the widespread assumption that clones are genetically uninteresting because they can't swap DNA through sex.

Study co-author Eric Karlin, a plant ecologist at Ramapo College in New Jersey, wrote, "You would expect one founding plant to have this rare trait. However, it is unlikely that there were many founding plants with each one having the same rare trait."

Surprisingly Diverse

Fossilized *S. palustre* moss remains have been found in 23,900-year-old peat near the summit of Kohala Mountain on Big Island (Hawaii). From these remains, Karlin and colleagues inferred that the moss had been in Hawaii at least that long, and perhaps longer.

The team analyzed the genetic diversity of the current population of moss on the island and determined a mutation rate. Using this rate, they estimated how long it took for the different moss populations to get to where they are genetically—about 50,000 years. The analysis also revealed surprising diversity—challenging the popular assumption that clones are genetically drab because they can't swap DNA through sex.

"They're not identical because mutations are always occurring," Karlin said.

Attack of the Clones

Given the absence of sex, the moss has been likely "trapped" on the Kohala summit, Karlin said. Sexual reproduction—which creates airborne spores—would be required for the plant to move elsewhere. But people have also lent a helping—though unwitting—hand in the moss's spread.

In the past century, people have used the moss for packing material, in doing so moving the species across Big Island as well as on the island of Oahu. "The peat moss has had explosive growth where it was introduced, especially on Oahu," Karlin said.

The moss's success comes at the cost of other local plants, however. "It's a problem," he said. "The moss completely changes the ecology of the ground layer. Instead of there being soil, there is a solid carpet of moss, and the seeds of many of the local plants don't grow in the moss layer." ∎

TRUTH: *SPHAGNUM* IS A GREEK NAME FOR A PLANT HISTORIANS HAVEN'T YET IDENTIFIED. *PALUSTRE* IS A LATIN WORD MEANING "MARSHY," OR "GROWING IN A MARSH."

World's Biggest Cave
Found in Vietnam

A massive cave uncovered in a remote Vietnamese jungle is the largest single cave passage yet found.

In 2009, a team of explorers measured the truly gigantic Son Doong cave to determine that it was indeed the world's biggest single cave passage. The cave's entrance had been discovered years earlier, but no one knew just how big a find it would turn out to be.

A New World Record

At 262-by-262 feet (80-by-80 meters) in most places, the Son Doong cave beats out the previous world-record holder, Deer Cave in the Malaysian section of the island of Borneo. Deer Cave is no less than 300-by-300 feet (91-by-91 meters), but it's only about a mile (1.6 kilometers) long. By contrast, explorers walked 2.8 miles (4.5 kilometers) into Son Doong, in Phong Nha-Ke Bang National Park, before being blocked by seasonal floodwaters—and they think that the passage is even longer.

> **TRUTH:**
> **A JUNGLE GROWS INSIDE VIETNAM'S SON DOONG CAVE.**

In addition, for a couple of miles Son Doong reaches more than 460-by-460 feet (140-by-140 meters), said Adam Spillane, a member of the British Cave Research Association expedition

A rock formation shines beneath a skylight in the Son Doong cave.

that explored the massive cavern. Spillane was in the first of two groups to enter the cave. His team followed the passage as far as a 46-foot-high (14-meter-high) wall.

"The second team that went in got flooded out," he said. "We're going back next year to climb that wall and explore the cave further."

Laser Precision

A local farmer, who had found the entrance to the Son Doong cave several years before, led the joint British-Vietnamese expedition team to the cavern in April 2009. The team found an underground river running through the first 1.6 miles (2.5 kilometers) of the limestone cavern, as well as giant stalagmites more than 230 feet (70 meters) high.

> **TRUTH:**
> THE SON DOONG CAVE BELONGS TO A NETWORK OF ABOUT 150 CAVES, MANY OF WHICH HAVE YET TO BE SURVEYED, IN THE ANNAMITE MOUNTAINS.

The explorers surveyed Son Doong's size using laser-based measuring devices. Such modern technology allows caves to be measured to the nearest millimeter, said Andy Eavis, president of the International Union of Speleology, the world caving authority, based in France. "With these laser-measuring devices, the cave sizes are dead accurate," he said. "It tends to make the caves smaller, because years ago we were estimating, and we tended to overestimate."

Eavis, who wasn't involved in the survey, agreed that the new findings confirm Son Doong's record status—despite the fact that he had discovered Borneo's now demoted Deer Cave. "This one in Vietnam is bigger," Eavis conceded.

However, the British caver can still claim the discovery of the world's largest cave chamber, Sarawak Chamber, also in Borneo. "That is so large it may not actually be beaten," he said. "It's three times the size of Wembley Stadium" in London.

Noisy and Intimidating

Son Doong had somehow escaped detection during previous British caving expeditions to the region, which is rich in limestone grottos. "The terrain in that area of Vietnam is very difficult," said expedition team member Spillane.

"The cave is very far out of the way. It's totally covered in jungle, and you can't see anything on Google Earth," he added, referring to the free 3-D globe software.

"You've got to be very close to the cave to find it," Spillane said. "Certainly, on previous expeditions, people have passed within a few hundred meters of the entrance without finding it."

The team was told that local people had known of the cave but were too scared to delve inside. "It has a very loud draft and you can hear the river from the cave entrance, so it is very noisy and intimidating," Spillane said.

Of more concern to the caving team were the poisonous centipedes that live in Son Doong. The explorers also spotted monkeys entering through the roof of the cave to feed on snails, according to Spillane.

"There are a couple of skylights about 300 meters [985 feet] above," he said. "The monkeys are obviously able to climb in and out." Planned return visits to the cave include bringing a biologist along to survey the cave's subterranean wildlife.

> **"Couldn't have done it without him . . . It took three expeditions to find Hang Son Doong. Khanh had found the entrance as a boy but had forgotten where it was. He only found it again last year."**
> **Howard Limbert**
> *British caver, on the help provided by a villager named Ho Khanh in discovering Son Doong*

Bigger Caves Waiting?

Eavis, of the International Union of Speleology, added that there are almost certainly bigger cave passages awaiting discovery around the world. "That's the fantastic thing about caving," he said. Satellite images hint, for example, that caves even larger than Son Doong lie deep in the Amazon rain forest, he said. ∎

New "Porta Potti" Flower Discovered

A rose by any other name would smell as sweet. And a new species of *Amorphophallus*— the genus that includes the "corpse flower"— still smells like rotting meat and feces.

Discovered on an island off the coast of Madagascar, the new-found plant grows up to 5 feet (1.5 meters) high and blooms once a year with a "really foul" stench, according to discoverer Greg Wahlert, a postdoctoral researcher in botany at the University of Utah.

The plant smells of rotting meat.

Lynn Bohs, a biology professor in the same lab as Wahlert, described the smell in a statement as a combination of "rotting roadkill" and a "Porta Potti." The new flower adds to the roughly 170 species in the *Amorphophallus* genus, which means "misshapen penis" in Greek after the phallic shape of the plants' flowers.

Stroke of Luck

Wahlert discovered the new species—named *A. perrieri*—in full bloom while collecting violets in two remote islands northwest of Madagascar in 2006 and 2007. Suspecting the plant might be a new species, he brought back samples and began cultivating them. After consulting with an *Amorphophallus* expert in the Netherlands, he confirmed that *A. perrieri* was a previously undescribed species.

Because *A. perrieri* is dormant for much of the year, Wahlert's discovery is a case of good timing. For months out of the year, there's little rain in that part of Madagascar, so the plants remain dormant underground. "These things are growing out of the most miserable soil," said Wahlert, who is working on a scientific paper about the species.

The specimen he is cultivating in the university's greenhouse shot up its flower in just two weeks. If Wahlert had been visiting the islands at a different time, "I could have very easily missed it."

TRUTH: CORPSE FLOWERS, MEMBERS OF THE *AMORPHOPHALLUS* GENUS, ATTRACT NOCTURNAL INSECTS SUCH AS BEETLES AND FLIES THAT USUALLY LAY EGGS IN ROTTING FLESH.

Stinky Flowers "Fascinating"

All *Amorphophallus* species emit smells to attract flies and other insects. Though a few emit more pleasant aromas, such as chocolate or spices, most smell terrible to human noses, Wahlert said. "You can imagine in Africa, where big game will die and rot in the sun . . . that's what they smell like," Wahlert said.

Despite the stench, he added, "I'm glad I got a stinky one. It's fascinating to me." ■

CHAPTER 10

Prehis
Life

Judging by the number of freaky fossils they uncover, you would think that paleontologists have a penchant for the peculiar. Their day jobs consist of finding strange beasts like the prehistoric snake longer than a school bus, rare ancient saber-toothed squirrels, and the massive "thunder thighs" dinosaur. Sea monsters and giant rabbits may seem weird to you and me, but they are par for the course for anyone who digs up the very distant past.

T. Rex, Other Dinosaurs Had Heads Full of Air

Dinosaurs were airheads—and not just because they had tiny brains, a new study says.

Three-dimensional scans of the skulls of *Tyrannosaurus rex* and other dinosaurs reveal the creatures had more empty space inside their heads than previously thought. These air spaces made the skulls light but strong and could have helped dinosaurs breathe, communicate, and hunt.

The extra room may even have paved the way for flight in some species. "Air is a neglected system that is actually an important contributor to what animals do," said study co-author Lawrence Witmer, a paleontologist at Ohio University in Athens.

Muscle Heads

Witmer and colleague Ryan Ridgely made detailed CT scans of air cavities in the skulls of two predators, *T. rex* and *Majungasaurus;* and two ankylosaurs,

> **TRUTH:**
> EVIDENCE SUGGESTS THAT *TYRANNOSAURUS REX* COULD EAT UP TO 500 POUNDS (230 KILOGRAMS) OF MEAT IN ONE BITE.

Panoplosaurus and *Euoplocephalus,* both plant-eaters with armored bodies and short snouts.

The results mark the first time scientists were able to accurately estimate the weight of a dinosaur's head. A *T. rex* head, for example, would have weighed more than 1,100 pounds (about 500 kilograms)—close to the average weight of an adult cow, Witmer and colleagues found.

Until now, paleontologists had to make do with estimates for the weight of dinosaur heads, said Tom Holtz, a paleontologist at the University of Maryland who was not involved in the research. "Larry's team is able to calculate a volume for the skull, so they can constrain the weight far more securely," Holtz said. "This is the next best thing to having a fleshy *T. rex* head to dissect."

Witmer estimates that *T. rex*'s head would have been 18 percent heavier if not for the air spaces in its skull. This savings may have allowed *T. rex* to pack more muscle onto its head, which possibly strengthened its bite and allowed it to tackle bigger prey.

"Crazy Straws"

The nasal airways in the ankylosaurs, however, were surprisingly convoluted. It was as if "crazy straws" had been rammed up the creatures' snouts, Witmer said. These winding airways were often located next to large blood vessels. "Whenever we see that, it raises the possibility that we're looking at heat transfer," Witmer said.

This setup would have allowed hot blood circulating through the creatures' heads to dump excess heat into the airways, helping to cool their brains and the rest of their bodies. The transferred heat also could have warmed up air the dinosaurs breathed, making gas exchange in the lungs easier.

In addition, the twisty nasal passages may have acted as resonating chambers for sounds. The two ankylosaur species examined had slightly different airways, so their voices would have been subtly different, Witmer said.

Hollow Bones

The research could provide new clues about how dinosaurs achieved flight. Some of the new study's research subjects were theropods, the group of dinosaurs from which modern birds are descended.

"Very often people have thought that birds have hollow bones because they fly, but it could be the other way around," Witmer said. "They could have evolved hollow bones for other reasons, and that gave them the lower body mass necessary to take to the air."

Hans-Dieter Sues is a dinosaur expert at the Smithsonian National Museum of Natural History in Washington, D.C., who did not participate in the research. Witmer "certainly makes a strong case for paranasal sinuses [air-filled spaces within the skull] reducing the weight of the skull in certain dinosaurs," Sues said. Sues cautioned, however, that "such functional hypotheses are difficult to test even in living species, including our own." ∎

> **"I've been looking at sinuses for a long time, and indeed people would kid me about studying nothing—looking at the empty spaces in the skull. But what's emerged is that these air spaces have certain properties and functions."**
>
> **Lawrence Witmer**
> *Chang Professor of Paleontology,
> Ohio University College of
> Osteopathic Medicine*

Massive "Sea Monster" Skull Revealed

Packing what may be the world's biggest bite, a recently revealed "sea monster" would have given Jaws a run for its money.

A scary new "sea monster" skull went on display in 2011 at the United Kingdom's Dorset County Museum, showing the world another of the terrifying predators that used to swim in the world's oceans.

The 7.9-foot-long (2.4-meter-long) skull belonged to a pliosaur, a type of plesiosaur that had a short neck, a huge, crocodile-like head, and razor-sharp teeth. When alive about 155 million years ago, the seagoing creature would have had a strong enough bite to snap a car in half, according to the museum.

The Dorset County Museum exhibit includes a life-size model of the pliosaur head to show what the animal would have looked like. A digital model of the Dorset pliosaur was also created using data from a high-energy, microfocus CT scanner. Scientists from several universities are teaming up for further research from the find, such as searching for fossil plankton that may have been preserved in mud surrounding the fossil pieces.

Mexican Pliosaur

Two German paleontologists found the 120-million-year-old pliosaur specimen—with a head the size of a car—in Mexico in 2002.

Fossil Find

Amateur collector Kevan Sheehan found the skull in pieces between 2003 and 2008 at the Jurassic Coast World Heritage Site, a 95-mile (152-kilometer)

SEA MONSTERS!

1: *Paleontologist Richard Forrest measures the **jawbone of the Dorset pliosaur** in 2009, when the find was first announced.*

2: *A life-size model of the **pliosaur head** from the Dorset County Museum exhibit.*

3: This skull belonged to a pliosaur, *a type of plesiosaur that had a short neck; a huge, crocodile-like head; and razor-sharp teeth.*

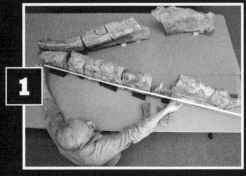

4: To aid in capturing prey, *some pliosaurs evolved features such as supersize eyes, fearsome teeth, or extremely long necks, according to* National Geographic *magazine.*

stretch of fossil-rich coastline in England. Sheehan gathered the pieces as they were washed out of a landslide on the coast of Weymouth Bay—the largest segment being more than 176 pounds (80 kilograms). Three pieces were later found by two other collectors, making the skull more than 95 percent complete, according to the museum.

Arctic Sea Monster

Found in Norway in 2008, one fossilized skull belonged to a 50-foot-long pliosaur that was longer than a humpback whale and had teeth the size of cucumbers.

"It is an amazing achievement to have recovered this fossil from an eroding cliff over such a long period of time and without losing any important pieces," Richard Edmonds, earth science manager for the Jurassic Coast Team, said in a statement. The team is a group of county officials who work to preserve the heritage area.

The Dorset County Council's museums service purchased the fossil, and later research by University of Southampton scientists suggests that it's the largest complete pliosaur skull ever found.

Yet Hans-Dieter Sues, a paleontologist at the National Museum of Natural History in Washington, D.C., cautioned in an email that it's too early to say if the skull is indeed the largest. "Some pliosaurs are gigantic animals, and there is an unfortunate tendency to brand every new find as the largest," said Sues. "However, no evidence is ever presented to support these claims, which make for good media coverage but are scientifically unwarranted."

A Different Species?

Pliosaurs were successful predators and evolved features to aid in capturing prey. Some pliosaur species developed supersize eyes, fearsome teeth, or extremely long necks. As for the Dorset specimen, further research—including CT scans—may show whether the 52-foot-long (16-meter-long) beast is a new species. Yet Sues noted that "the classification of late Jurassic pliosaurs is still a mess, and, in absence of a reliable, published, modern review of all European specimens, it is hard to sustain any claim of a new species." ■

Biggest Snake Discovered

Was Longer Than a Bus

Steamy year-round temperatures may have created the perfect environment for the world's biggest snakes. Could climate change bring their kind back?

The world's biggest snake was a massive anaconda-like beast that slithered through steamy tropical rain forests about 60 million years ago, says a new study that describes the ancient giant.

How Big Was He?

Fossils found in northeastern Colombia's Cerrejon coal mine indicate the reptile, dubbed *Titanoboa cerrejonesis,* was at least 42 feet (13 meters) long and weighed 2,500 pounds (1,135 kilograms).

"That's longer than a city bus and . . . heavier than a car," said lead study author Jason Head, a fossil-snake expert at the University of Toronto, Mississauga in Canada and a research associate with the Smithsonian Institution. Previously the biggest snake known was *Gigantophis garstini,* which was 36 to 38 feet (11 to 11.6 meters) long. That snake lived in North Africa about 40 million years ago.

> # TRUTH:
> ## THE *TITANOBOA CERREJONENSIS* WAS LONGER THAN A CITY BUS AND HEAVIER THAN A CAR.

The world's biggest snake lounges in the tropics in this illustration.

Hans-Dieter Sues, associate director for research and collections at the Smithsonian's National Museum of Natural History, was not involved with the study but has seen the snake fossils. Sues noted that humans would stand no chance against one of these giants, which killed their prey by slow suffocation.

"Given the sheer size—the sheer cross-section of that snake—it would be probably like one of those devices they use to crush old cars in a junkyard," Sues said.

Biggest Snake Needed the Heat

In addition, the snake's heft indicates that it lived when the tropics were much warmer than they are today, a find that holds potential implications for theories of once and future climate change.

Scientists know there's a link between a snake's body size, how fast it uses and produces energy, and climate. "We were able to use the snake, if you will, as a giant fossil thermometer," study author Head said.

His team found that, for *Titanoboa* to reach its epic proportions, it would mean year-round temperatures would have been about 90 degrees Fahrenheit (32 degrees Celsius)—significantly hotter than today's tropics.

This supports the idea that tropical temperatures spike as the rest of the world heats up due to global warming, the study authors say. The competing theory is that, during bouts of warming, the tropics stay about the same average temperature as they are today while areas north and south of the Equator heat up.

James Zachos, an expert on ancient climates at the University of California, Santa Cruz, who was not involved in the study, agreed. As the biggest known snake, *Titanoboa* supports the idea of "much hotter tropics during extreme greenhouse periods," Zachos said.

The Return of the Giant Snakes?

Study co-author Jonathan Bloch is a vertebrate paleontologist at the University of Florida's Museum of Natural History in Gainesville. The same Colombian coal mine that contained the biggest snake also yielded massive turtles and crocodiles, he said.

"You can think about it as an ecosystem dominated by giants, I think, and these are probably giants that got large because of the warmer mean annual temperature," he said.

The findings paint a picture of what the future might hold if supercharged global warming takes place. According to some models, global temperatures could approach the same levels that gave rise to the biggest snake by the end of this century.

If current greenhouse gas emissions continue apace, there's a chance snakes the size of *Titanoboa* could return, Bloch said. "Or maybe snakes would go extinct in the tropics," he said. "In other words, the warming could happen so rapidly that they wouldn't have time to adapt." ∎

Opposite End of the Spectrum

The world's smallest snake, the *Leptotyphlops carlae,* is less than four inches long and was discovered in 2008 on the Caribbean island of Barbados. Due to its tiny size, it can easily be mistaken for an earthworm. *L. carlae* is a new species of snake that belongs to a group called thread snakes, and it may be on the verge of extinction.

Oldest Animal Discovered

Aquatic African Ancestor

**Mirror, mirror on the wall.
Who is the earliest ancestor of us all?
The answer might surprise you.**

Microscopic, spongelike African fossils could be the earliest known animals—and possibly our earliest evolutionary ancestors, scientists say. The creature, *Otavia antiqua,* was found in 760-million-year-old rock in Namibia and was as tiny as it may be important.

> **TRUTH:**
> **THERE ARE 9,000 SPECIES OF SPONGES, AND THEY OCCUPY VIRTUALLY EVERY AQUATIC HABITAT ON EARTH.**

Starting Small

"The fossils are small, about the size of a grain of sand, and we have found many hundreds of them," said study leader Anthony Prave, a geologist at the University of St. Andrews in the United Kingdom. "In fact, when we look at thin sections of the rocks, certain samples would likely yield thousands of specimens. Thus, it is possible that the organisms were very abundant."

From these tiny "sponges" sprang very big things, the authors suggest. As possibly the first multicellular animals, *Otavia* could well be the forerunner of dinosaurs, humans—basically everything we think of as "animal."

A modern-day stove-pipe sponge

Built to Last

Prior to the new discovery, the previous earliest known "metazoan"—animals with cells differentiated into tissues and organs—was another primitive sponge, dated to about 650 million years ago.

Based on where the new fossils were found, Prave and his colleagues think *Otavia* lived in calm waters, including lagoons and other shallow environments. The team thinks *Otavia* fed on algae and bacteria, which the animal drew through pores on its tubelike body into a central space. There the food was digested and absorbed directly into *Otavia*'s cells. The simple setup seems to have worked.

The fossil record indicates *Otavia* survived at least two long-term, severe cold snaps known as "snowball Earth" events, when the planet was almost completely covered in ice. Despite such wild environmental swings, "the oldest and youngest *Otavia* fossils all have the same quasi-ovid form, with large openings leading from the exterior," Prave said in an email.

In short, the animals didn't evolve much, he said—suggesting that, at least for its roughly 200 million years of existence, *Otavia* was built to last. ■

Ancient "Saber-Toothed Squirrel" Found

Fossils of ancient mammals are very rare—mostly because of their small size. But this latest find, a squirrel with very sharp teeth, gives paleontologists new insight into an ancient world.

The fossilized skull and teeth of a fanged, shrewlike mammal have been found in Argentina, a new study says. The new species—dubbed *Cronopio dentiacutus* for its narrow snout and long fangs—was about 8 to 9 inches (20 to 23 centimeters) long and likely used its pointy teeth to hunt and eat insects.

The Little Guys

C. dentiacutus was very small, like most mammals at the time. "These were the tiny little guys that would squirrel in between the toes of the dinosaurs trying not to get stepped on," says Guillermo Rougier, a paleobiologist who specializes in these animals. Mammals didn't grow to be the size of cats and dogs until after the larger dinosaurs became extinct.

Big Teeth

The second oldest mammal skull ever recovered from South America, *C. dentiacutus* existed when dinosaurs still roamed Earth. Paleontologists found the mostly complete skull in 2002 outside a rural village in northern Argentina. At the time, however,

This is an artist's rendition of what the saber-toothed squirrel might have looked like.

the skull was mostly hidden in rock, and its identity remained a mystery.

So in 2005 the scientists sent the skull to a technician, who spent three years removing the rock from around the fossil—finally revealing a saber-toothed, squirrel-like creature.

"When [the movie] *Ice Age* came out, we thought the squirrel character in it looked ridiculous, but then we found something like it," said study leader Guillermo Rougier, a paleontologist at the University of Louisville in Kentucky. "This animal looks very peculiar, with long snout and canines, and it highlights that we know so little. Surprises like this are bound to happen."

Rare Remains

Both mammals and dinosaurs appeared near the end of the Triassic period, some 220 million years ago. When dinosaurs disappeared about 65 million years ago, mammals thrived. But ancient-mammal fossils are still exceedingly rare, mostly because of their small sizes.

"Getting on your hands and knees . . . is how you find small-mammal fossils," Rougier said. "It is not very glamorous. You basically roll in the dirt all day." As a result, paleontologists know of roughly one genus of mammal for every million years between 65 million and 220 million years ago—making for a woefully incomplete record. "Imagine trying to reconstruct the history of life with that information," Rougier said. "We're certain there were hundreds of [genuses], but for now it's like trying to reconstruct the brilliance of James Joyce with just ten of his words." ■

Prehistoric Bird Had Wings Like Nunchucks

Don't make these ancient flightless birds angry. You wouldn't like them when they're angry.

A flightless bird with wings like martial arts weapons once thrashed its foes on what's now Jamaica, a new study says.

Dubbed *Xenicibis,* the prehistoric bird wielded its unusual wings like nunchucks, or nunchakus, swinging its upper arms so that thick, curved hand bones hinged at the wrist would deliver punishing blows. The weaponlike wings are so unique that study co-author Nicholas Longrich of Yale University at first assumed the odd limbs were evidence of a deformity.

"There are a lot of birds that do have weaponry," Longrich said. "They just don't have anything like this."

Wings That Bear Arms

Xenicibis is an extinct member of the ibis family that grew to be the size of a large chicken. Although the ancient animal had been known for

> "No animal has ever evolved anything quite like this. We don't know of any other species that uses its body like a flail. It's the most specialized weaponry of any bird I've ever seen."
>
> **Nicholas Longrich**
> *research leader, Yale University*

years, new analysis of several recently discovered partial skeletons has only now revealed the odd wing bones.

Wings are among the most powerful parts of birds, and some modern species boast specialized wing weapons. Screamer birds, for instance, use daggerlike spurs to fight for mates, while steamer ducks have wrist knobs that can break bones or even kill other birds. Since *Xenicibis* was flightless, Longrich and colleagues think the ancient bird evolved to have wings that were even more specialized for combat.

Weapon Wings

Examinations of the skeletons uncovered evidence of past violence—one wing had a fractured hand bone, and another had a centimeter-thick upper-arm bone that had been cleanly snapped in half. The bird's living ibis relatives are anatomically similar except for the "weapon wings," so modern ibis behavior may shed light on when and how *Xenicibis* used its built-in arsenal. Male ibis, for example, regularly battle each other over territories for nesting and feeding.

But *Xenicibis* may also have needed its powerful wings to fight off other threats. "Most flightless birds, like the dodo, had absolutely no predators around. But Jamaica had many snakes, raptors, and other potential predators," Longrich said. "So maybe they just needed a bit more defense."

Boxing Birds

The prehistoric *Phorusrhacid*, also known as the terror bird, had a hatchet-like skull that was used to kill prey with a succession of punishing blows. The skull was the size of a modern-day horse's head, and was very strong and rigid when driven straight down into prey. Like the *Xenicibis*, the terror bird was also flightless, so it most likely evolved to have such a skull in order to survive on land.

Nunchucks for Dinner?

Longrich and colleagues are most curious about the ultimate fate of the prehistoric birds. Fossils indicate *Xenicibis* lived at least as recently as 12,000 years ago. But the record is too sparse at this point to know exactly when they vanished—or if ancient humans might have been involved.

"Did they disappear long before people arrived in Jamaica [several thousand years ago] or last long enough for people to wipe them out?" Longrich said. "They were a convenient size for a family dinner and couldn't fly. Humans are pretty handy with clubs also, so it might not have been a very fair fight. But at this point we just don't know." ∎

"Thunder Thighs" Dinosaur
Thrashed Predators to Death?

Ill-tempered dinosaur used her massive muscular legs to kick away fearsome raptors, expert says.

A newfound dinosaur species that used its "exceptionally powerful" thighs to kick predators likely had a bad temper to boot, one expert says.

The 46-foot-long (14-meter-long) *Brontomerus mcintoshi* had an immense blade on its hipbones, where strong muscles would have attached, according to a new study. "These things don't happen by accident—this is something that's clearly functional," said study co-author Mathew Wedel.

Thunder Thighs Attack

The team suspects the dinosaur—a type of sauropod, or plant-eating, four-legged lumberer—used its massive legs to either maneuver over hilly ground or deliver "good, hard" kicks to predators, said Wedel, assistant professor of anatomy at Western University of Health Sciences in Pomona, California.

Brontomerus—"thunder thighs" in Greek—may have even attacked like a modern-day chicken, relentlessly kicking and stomping pursuers to death, he added. "I could only imagine how ill-tempered these sauropods would have been," Wedel said—as are most birds, dinosaurs' modern-day descendants.

> **TRUTH:**
> **THE LARGEST DINOSAURS WERE VEGETARIANS.**

A *"thunder thighs"* dinosaur mother protects her offspring in an artist's conception.

In both cases, "you've got a little brain, you're permanently paranoid about all these meat-eaters around, and you're trying to protect your young."

Big Leg Muscles

Thunder thighs' bones were first found in 1994, when scientists rescued two partial skeletons of the then unidentified dinosaur from a fossil quarry that had otherwise been looted in eastern Utah.

When Wedel and colleagues examined the bones in 2007, they realized they'd found a new species—and an "extreme" one at that, Wedel said. For instance, the shapes of the newfound species' bones showed it had the largest leg muscles of any sauropod yet found.

B. mcintoshi likely needed such extreme defenses to fight off "terrifying" predators such as *Deinonychus* and *Utahraptor*, raptors that lived alongside the plant-eater about 110 million years ago in the early Cretaceous period, he said.

Sauropod Safari

The prehistoric animals roamed a landscape that would have resembled Africa's Serengeti, laced with rivers and mud holes and distinguished by vast, dry upland areas, Wedel noted. Herds of cowlike plant-eaters called *Tenontosaurus* would have dotted the plains.

"If I could shoot you back in a time machine, it would have been like going on safari, except you'd want something more robust than a Land Rover—maybe a tank," he said. "The sauropods were probably beautiful animals if you were a long way away with binoculars," he added.

"But up close, [they were] probably a nightmare." ∎

Giant Bugs Once Roamed the Earth

Predatory dragonflies the size of modern seagulls ruled the air 300 million years ago, and it's long been a mystery how these and other bugs grew so huge.

Dragonflies as big as seagulls? Oversized giant cockroaches? Why did these bugs get so big? To avoid oxygen overdose, one study hints.

Too Much of a Good Thing

The leading theory is that ancient bugs got big because they benefited from a surplus of oxygen in Earth's atmosphere. But a new study suggests it's possible to get too much of a good thing: Young insects had to grow larger to avoid oxygen poisoning.

"We think it's not just because oxygen affects the adults but because oxygen has a bigger effect on larvae," said study co-author Wilco Verberk of Plymouth University in the United Kingdom. "So a larval perspective might lead to a better understanding of why these animals existed in the first place, and maybe why they disappeared."

Baby Bugs Can't Control Their Gases

Fossils show that giant dragonflies and huge cockroaches were common during the Carboniferous period, which lasted from about 359 to 299 million years ago. During this time, the rise of vast lowland swamp forests led to atmospheric oxygen levels of around 30 percent—close to 50 percent higher than current levels.

According to previous theories about insect gigantism, this rich oxygen environment allowed adult bugs to grow to ever larger sizes while still meeting their energy needs.

For the new study, Verberk and colleague David Bilton instead focused on how varying oxygen levels affect stonefly larvae, which, like dragonflies, live in water before becoming terrestrial adults. Higher concentrations of oxygen in air would have meant higher concentrations dissolved in water.

The results showed that juvenile stoneflies are more sensitive to oxygen fluctuations than their adult counterparts living on land. This may be because insect larvae typically absorb oxygen directly through their skin, so they have little or no control over exactly how much of the gas they take in. By contrast, adult insects can regulate their oxygen intake by opening or closing valvelike holes in their bodies called spiracles.

While crucial for life, oxygen can be poisonous in large quantities: Humans exposed to excess oxygen can suffer cell damage leading to vision problems, difficulty breathing, nausea, and convulsions. It's likely the larvae of many ancient insects also passively absorbed oxygen from water and were not able to regulate their oxygen intake very well—a big danger when oxygen levels were so high.

One way to decrease the risk of oxygen toxicity would have been to grow bigger, since large larvae would absorb lower percentages of the gas, relative to their body sizes, than small larvae. "If you grow larger, your surface area decreases relative to your volume," Verberk explained.

Lower Oxygen Levels

The new theory could also explain why giant insects continued to exist even after Earth's atmospheric oxygen levels began decreasing, he said. "If oxygen actively drove increases in body mass to avoid toxicity, lower levels would not be immediately fatal, although in time, they [would] probably diminish performance of the larger insects," since adults would have evolved to require more oxygen and would get sluggish in air with lower levels, Verberk said.

"Such reduced performance will eventually have made it possible for other species to outcompete the giants." ∎

TRUTH: SEVENTY-FIVE PERCENT OF ALL ANIMALS ARE INSECTS.

Brainy Birds

Out-Thought Doomed Dinosaurs?

Having a bird brain wasn't such a bad thing 65 million years ago. It was these bigger bird brains that kept them alive while the dinosaurs and their teeny brains went extinct.

Birds survived the global catastrophe that wiped out their dinosaur relatives due to superior brainpower, a 2009 study suggests.

Bird Brains = Big Brains

A pair of prehistoric seabirds found in southeast England by Victorian-era fossil hunters were examined by researchers from the Natural History Museum in London. The two 55-million-year-old skulls suggest the ancestors of modern birds developed larger, more complex brains than previously thought.

This implies that bird ancestors had a mental edge over non-birdlike dinos and flying reptiles, so they were better able to adapt after the so-called K-T mass extinction event around 65 million years ago, said study co-author Angela Milner.

Some ancient groups of birds did go extinct, she noted, so it wasn't feathers or warm-bloodedness that gave modern birds a leg up. "It had to be something else," she said, "and it seems to be this bigger brain."

> **TRUTH:**
> RESEARCHERS HAVE FOUND THAT PRE-HISTORIC SEABIRDS' BRAINS WERE NEARLY THE SAME SIZE AS THOSE IN BIRDS ALIVE TODAY.

Advantage, Birds

The study, published in 2009 in the *Zoological Journal of the Linnean Society*, was based on two specimens from the Natural History Museum's vast fossil collection. *Odontopteryx toliapica* belonged to an extinct group of giant, bony-toothed seabirds, while *Prophaethon shrubsolei* was a prehistoric relative of ternlike tropical seabirds.

Milner and colleagues used CT scans of the skulls to make models of the size and shape of the fossil birds' brains. What they found is that the ancient bird brains were almost the same size as those in birds alive today. The older noggins also showed early growth of a brain region known as the wulst.

"It seems to be the area that's involved in more complex behavior and cognition, such as being able to learn about your environment and remember it," Milner said. So after the K-T event, she said, these birds "were just better equipped to deal with challenging physical conditions."

Fossil bird skulls that have not been flattened out over time are extremely rare, and no examples are known from the time of the K-T event. But Milner says the brain advances seen in the 55-million-year-old birds would probably have begun more than 65 million years ago.

And fossils of the oldest known bird, *Archaeopteryx*, which lived 147 million years ago, reveal its brain was "nowhere near as well developed as the ones we looked at," she said.

Desperately Seeking Fossils

Julia Clarke, a geoscientist at the University of Texas at Austin who was not involved with the study, says there are various competing theories to explain why birds outlived the dinosaurs. One idea is that the ancestors of all living birds came from the southernmost part of the southern supercontinent Gondwana, where they escaped the worst of the environmental fallout from the K-T event.

Another theory is that modern bird lines evolved in coastal habitats that also were less heavily impacted. As well as providing valuable new evidence for the evolution of birds, she said, the latest study offers an intriguing new theory that will motivate paleontologists to look harder and farther to find more fossils.

"We still desperately need good fossils sampling brain and skeletal features in the species that are very close but outside the [evolutionary tree] of all living birds," Clarke said. "We can only get so close to understanding the brains of the earliest birds with the sample of known species currently available." ∎

Prehistoric "Shield"-Headed Croc Found

The prehistoric "ShieldCroc" had a fierce look, but weak jaws. Paleontologists say this "one of a kind" creature was more pelican than pro-wrestler when it hunted.

A new prehistoric croc sporting an odd head "shield" has been found in Morocco, paleontologists say. Dubbed ShieldCroc, the animal's head appendage was surrounded by blood vessels and covered with a sheath like those seen in frilled dinosaurs, including *Triceratops*.

Scoop Like a Pelican

At 30 to 35 feet (9 to 11 meters) long, the river-dwelling monster would have preyed on other giant animals of the late Cretaceous, such as 13-foot-long (4-meter-long) coelacanths. But ShieldCroc likely boasted relatively weak jaws, at least compared with those of today's crocodiles.

"It's fairly certain that it belonged to a group of crocodyliforms—including the flat-headed crocs—that had really thin, weak jaws and weak chin joints," said researcher Casey Holliday, a paleontologist at the University of Missouri. Crocodyliforms are part of a group known as the crocodilians, which includes modern-day alligators, caimans, and more.

> **TRUTH:**
> SCIENTISTS OFTEN USE THE HEAD SIZE OF AN ANIMAL TO ESTIMATE ITS TOTAL LENGTH.

"So they weren't wrestling dinosaurs on the water's edge. They would have been quick, snap feeders waiting for prey to come by and then grabbing it and swallowing it with large, basket-shaped mouths—something like a pelican would do," Holliday said.

Showy Headpiece

A piece of ShieldCroc's skull landed in Canada's Royal Ontario Museum in the early 2000s, but Holliday and colleagues have only recently studied the specimen and its odd headpiece. It's difficult to determine what purpose the shield served when the animal lived, some 99 million years ago, Holliday noted.

But after rigorous evaluation of the fossil and studies of comparative behaviors with modern crocodilians, scientists suggest the shield may have helped ShieldCroc regulate its temperature and communicate with other ShieldCrocs.

For instance, some crocodyliforms and living crocodilians, such as the Cuban crocodile, have horns on the sides of their heads, which males use to impress females and scare away other males. "We kind of see ShieldCroc having similar behaviors and showing off the roof of its head," Holliday said.

An illustration of ShieldCroc snatching its prey

Despite these possible similarities with modern crocodilians, the animal appears to have been one of a kind, said Christopher Brochu, a University of Iowa paleontologist, who wasn't involved in the study. "There's nothing quite like this among the birds or the crocodilians, which are the two closest living relatives of this thing."

Five Oddball Prehistoric Crocs

1. **RatCroc**—rodent-like, with buck teeth for rooting through the ground for tubers and simple animals
2. **PancakeCroc**—flat-bodied, lying motionless and waiting for prey to swim into its thin, 3-foot-long jaws
3. **DuckCroc**—used its long, smooth, sensitive nose to poke through vegetation and its hook-shaped teeth to capture frogs and small fish in shallow water
4. **DogCroc**—a plant-eater with lanky legs, which indicates it was quick enough to run into water if threatened
5. **BoarCroc**—a 20-foot-long "saber-toothed cat in armor" that preyed on dinosaurs with its three sets of fangs

Croc Evolution

ShieldCroc's discovery in Morocco could suggest that modern crocs evolved in what's now the Mediterranean—a theory that remains hotly debated among crocodilian experts. But there's no doubt the animal provides evidence of astonishing crocodyliform diversity in the Southern Hemisphere during the late Cretaceous, said Holliday, who described the new species at the 2011 Annual Meeting of the Society of Vertebrate Paleontology in Las Vegas.

"It definitely points to . . . Africa [as] a melting pot of different crocodyliforms living in the same region at the same time," Holliday said. "One lineage, including DogCroc, BoarCroc, and others tended to be terrestrial, while another group, including SuperCroc, were big, aquatic, predatory crocs. ShieldCroc represents another group and a more modern flavor of crocs."

Widespread Crocodyliforms

With the discoveries of ShieldCroc and related species, University of Iowa's Brochu said, "We're beginning to realize just how diverse and even bizarre the crocodyliforms were in the Southern Hemisphere," he said. "The group was extremely widespread, and in some places crocodyliforms may have been among the major predators and even herbivores. And in some places they really were simply bizarre." For instance, "in the southern Mediterranean, including North Africa, we're seeing these animals that look nothing like any living crocodilian." ∎

Largest Flying Bird

Could Barely Get off Ground, Fossils Show

Taking off was no easy task for an enormous bird that lived in the Andes mountains six million years ago. How did this big bird get airborne?

The largest bird that ever flew was an expert glider but was too heavy to fly by flapping its wings, researchers say.

Getting off the ground was a challenge for the 155-pound (70-kilogram) *Argentavis magnificens,* a condorlike bird that lived in the Andes mountains and the pampas of Argentina about six million years ago. Despite its massive flight muscles and 21-foot (6.4-meter) wingspan, the giant bird probably could not generate enough lift to take off from a level surface, according to a new study.

Excellent Glider

Like human hang gliders, *Argentavis* probably had to run downhill into a headwind to become airborne, said Sankar Chatterjee of Texas Tech University in Lubbock. "Takeoff capability is the limiting factor for the size of flying birds, and *Argentavis* almost reached the upper limit," Chatterjee said. "Heavier birds such as the ostrich had to give up flight." Once aloft, however, *Argentavis* was no ostrich. Despite weighing as much as 16 bald

Fly Like a Bird

NASA engineers have programmed a model airplane to look for rising columns of hot air called thermals and use them to soar like a bird, similar to the way the prehistoric *Argentavis magnificens* probably flew. This way of flight allows air currents to do most of the work required to gain altitude.

> "Mythological versions of giant soaring birds appear in religions all over the world. What we have done is shown that it would have been possible for a so-called monster bird to fly."
>
> **Sankar Chatterjee**
> *curator of paleontology,*
> *Texas Tech University*

eagles, Chatterjee said, "it was an excellent glider, like a sail plane."

Going Up

The new understanding of *Argentavis* flight comes from an unusual collaboration between paleontologists and a retired aeronautical engineer. The researchers took measurements from *Argentavis* fossils and then conducted their analysis using a computer program designed to study flight performance in helicopters.

"Birds are commonly compared with aircraft, but in reality helicopters are a better analogy," Chatterjee said. Unlike engine-powered airplanes, he noted, birds rely on their wings for both forward thrust and vertical lift, the two components necessary for flight.

Although *Argentavis* could not wing skyward on its own, the researchers say, it could have reached high altitudes by riding winds deflected upward over mountains. More commonly, particularly in open terrain, *Argentavis* probably gained elevation by circling inside rising columns of warm air, known as thermals.

The huge flyer may have traveled hundreds of miles by repeatedly riding

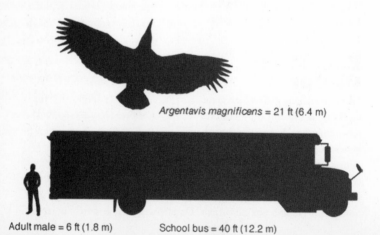

Argentavis magnificens = 21 ft (6.4 m)

Adult male = 6 ft (1.8 m)   School bus = 40 ft (12.2 m)

thermal "elevators" and then soaring gradually back to earth, Chatterjee said.

Some of the largest flying birds today, such as condors and eagles, pursue a similar strategy. Although capable of powered flight, these species save energy by letting air currents do most of the work required to gain altitude.

Predator or Scavenger?

In the past, researchers have disagreed as to whether *Argentavis* was a predator, like most hawks and eagles, or a scavenger. Chatterjee and co-author Kenneth Campbell, of the Natural History Museum of Los Angeles County, say fossil details indicate the species was an active predator.

"[The birds' skull] was adapted for catching prey and swallowing it whole," Campbell said. "Its jaw mechanics were not suited for tearing flesh from carcasses, as in vultures, nor for tearing prey animals apart for swallowing, as in eagles and owls."

But Paul Palmqvist, of the University of Malaga in Spain, has argued that a flying species as large as *Argentavis* must have been a scavenger. Palmqvist's argument is based in part on a predictable relationship between body size and foraging area seen in predatory hawks and eagles today.

Given its huge size, Palmqvist says, a predatory *Argentavis* would not have been able to cover enough ground and locate enough prey to meet its daily needs. "A vulturelike behavior is more reasonable, as vultures have smaller range areas," Palmqvist said. "Carrion is more available than living flesh."

The new flight analysis, he said, also tends to support his view. "Given its lack of maneuverability, a predator this size would have a problem landing on its prey," Palmqvist noted. But Chatterjee and Campbell said the species was certainly a capable enough flyer to attack live prey—probably rabbit-size mammals—from the air. ∎

> **TRUTH:**
> ONE OF THE LARGEST BIRDS TODAY IS THE ANDEAN CONDOR, WHICH HAS A WINGSPAN OF ABOUT 9 FEET AND WEIGHS 25 POUNDS.

Oldest Fossil Brain

Found in Kansas

Found in "bizarre" prehistoric fish, scientists have found a 300-million-year-old brain—the oldest fossil of its kind.

Digital x-ray images of a "bizarre" 300-million-year-old shark relative have revealed the oldest known fossilized brain, researchers announced in 2009.

Hard Brain Is a Rare Find

The unusual discovery raises hopes that scientists will find other ancient brains and use them to study how gray matter has evolved, said John Maisey, a paleontologist at the American Museum of Natural History in New York. "The brain…is remarkably soft tissue—brain tissue is mostly water," Maisey said. "To preserve anything is quite remarkable."

> **TRUTH:**
> THE BRAIN GROWTH OF MODERN SHARKS, RAYS, AND CHIMAERA FISH SLOWS AS THEY AGE, EVEN THOUGH THE REST OF THEIR BODIES CONTINUE TO EXPAND.

Ratfish Ancestor

The fossil was found in an iniopterygian, an extinct ancestor of modern ratfishes, also called "ghost sharks" or chimaeras. The fish are also distant relatives of sharks and rays. Maisey said the ancient fish, which swam in an ocean that once covered the midwestern United States,

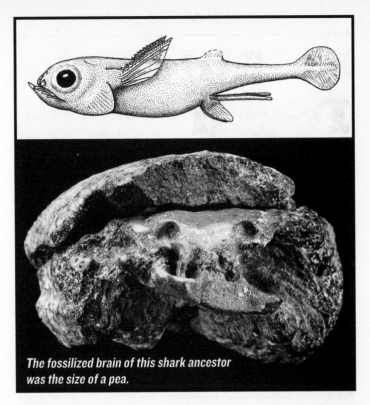

The fossilized brain of this shark ancestor was the size of a pea.

would have fit in the palm of a human hand. Despite their small size, the fish sported a strange appearance: huge eye sockets, rows of sharp sharklike teeth, tails with clubs, large pectoral fins, and spikes on the tips of their fins.

Fish Brain

The scans revealed the fish had a pea-size brain much smaller than the braincase itself. This is similar to modern sharks, rays, and chimaera fish, whose brain growth slows as they age, even as the rest of their bodies expand.

The iniopterygian's brain has a large lobe for vision, and the skull has relatively large eye sockets. This suggests the fish "was using its eyes as a major way to locate prey," Maisey said. In addition, the hearing-related section of the brain is flattened. This reflects the curious arrangement of the iniopterygian ear, which was optimized for side-to-side movement, but not up and down movement.

"It is really a very puzzling fish as to how it would have moved around and what it could have done," Maisey said. "They are really, really bizarre." ■

Did Love Make Neanderthals Extinct?

A new study claims that Neanderthals were done in by the more successful *Homo sapiens*. But extinction was not through acts of violence, but acts of love.

Neanderthals may have been victims of love, or at least of interspecies breeding with modern humans, according to a new study.

As the heavy-browed species ventured farther and farther to cope with climate change, they increasingly mated with our own species, giving rise to mixed-species humans, researchers suggest. Over generations of genetic mixing, the Neanderthal genome would have dissolved, absorbed into the *Homo sapiens* population, which was much larger.

A reconstruction of a Neanderthal female

"If you increase the mobility of the groups in the places where they live, you end up increasing the gene flow between the two different populations, until eventually one population disappears as a clearly defined group," said study co-author C. Michael Barton, an archaeologist at Arizona State University's School of Human Evolution and Social Change.

Doing What Comes Naturally

Some theories suggest Neanderthals disappeared about 30,000 years ago because the species wasn't able to adapt to a cooling world as well as *Homo sapiens*.

Barton tells a different tale, suggesting that Neanderthals reacted to the onset of the Ice Age the same ways modern humans did, by ranging farther for food and other resources. "As glaciation increased, there was likely less diversity in land use, so Neanderthals and modern humans alike focused on a particular survival strategy that we still see today at high latitudes," Barton said.

"They establish a home base and send out foraging parties to bring back resources. People move farther and have more opportunity to come into contact with other groups at greater distances. The archaeological record suggests that this became more and more common in Eurasia as we move toward full glaciation."

More frequent contact led to more frequent mating, the theory goes, as the two groups were forced to share the same dwindling resources. "Other things might have happened," Barton said. "But in science we try to find the simplest explanation for things. This theory doesn't include massive migrations or invasions—just people doing what they normally do."

To estimate the effects of the assumed uptick in interspecies mating, Barton's team conducted a computational modeling study that spanned 1,500 Neanderthal generations. In the end, the model results supported the not entirely new idea that Neanderthals were "genetically swamped" by modern humans.

> "Normally the first groups who [encounter] a new population are men, hunting parties perhaps. And men, being the way they are—if they meet women from another population, there is bound to be interbreeding."
>
> **Bence Viola**
> *paleoanthropologist, on possible reasons why Neanderthals and humans interbred*

"Genetic Swamping"

Though it's a relative underdog among Neanderthal-demise theories, genetic swamping is a well-known extinction cause among plant and animal species. A smallish group of native, localized trout, for example, may lose their genetic identity after a large influx of a different species with which the native fish are able to breed.

"When endemic populations are specialized, and for some reason there is a change in their interaction with adjacent populations, and that interaction level goes up, they tend to go extinct—especially if one population is much smaller than the other," Barton explained. "In conservation biology this is called extinction by hybridization."

TRUTH:
DNA EVIDENCE SUGGESTS THAT NEANDERTHALS MIGRATED AS FAR EAST AS SIBERIA.

On the Hunt

Paleoanthropologist Bence Viola said other models have produced different results, and some studies have concluded that relatively little interbreeding occurred. But Viola, of the Max Planck Institute for Evolutionary Anthropology in Leipzig, Germany, is intrigued by Barton's research.

"From an archaeological and anthropological perspective, this sounds interesting and closer to what I believe—that you can have a lot of interbreeding," Viola said. "Normally the first groups who [encounter] a new population are men, hunting parties perhaps. And men, being the way they are—if they meet women from another population, there is bound to be interbreeding."

Barton believes interbreeding caused other distinct human and human-ancestor groups to fade away. "But their genes didn't disappear," he added. "And their culture probably didn't disappear either but was blended into a larger population of hunter-gatherers."

The Max Planck Institute's Viola believes interbreeding was a cause—but not the cause. "Neanderthals disappeared around 30,000 years ago, and that was a period when the climate turned colder, and that likely made it physically harder for them to survive," Viola said. "They also may have been exposed to some type of disease that modern humans brought from Africa and for which they had no immunity.

"Of course these are all things that are very hard to study archaeologically," Viola added. "So these models are a great tool for investigating ideas." ∎

"Nasty" Little Predator

From Dinosaur Dawn Found

**Deadly and dog size, the dinosaur *Eodromaeus*
lived in Argentina 230 million years ago,
in the time before dinosaurs dominated.
What can we learn from the little monster?**

The new species is providing fresh insight into the era before
dinosaurs overtook other reptiles and ruled the world, a new fossil study says.
"This is the most complete picture we
have of a predatory dinosaur lineage—
what it looked like at the very beginning,"
said study co-author Paul Sereno. "It was
small but nasty—this animal was fast."

Nasty, Brutish, and Short

One of the earliest known dinosaurs,
Eodromaeus was only about 4 feet
(1.3 meters) long and would have barely
reached the knees of an adult human.
But this unassuming little dinosaur gave
rise to the theropods, including *Tyran-
nosaurus rex* and the "terrible claw," *Deinonychus,* the new study suggests.

Like those fearsome descendants, *Eodromaeus* had a long rigid tail, a
unique pelvis shape, and air sacs in its neck bones that may have been related

> "It was very cute;
> you'd want it as a
> pet. But it might
> be best as a guard
> dinosaur, to keep
> the dogs away."
>
> **Paul Sereno**
> *paleontologist and
> study author*

BIG FIND, LITTLE DINOSAUR

1 : *The **Valley of the Moon** in northwestern Argentina, where fossils of Eodromaeus and Eoraptor were found.*

2 : *A **replica skeleton** of Eodromaeus.*

3 : *Needle-like teeth spike a **full-size replica of the skull** of the new Eodromaeus dinosaur.*

4 : Eodromaeus *was a **distant relative** of the theropods, including Tyrannosaurus rex and the "terrible claw," Deinonychus.*

1

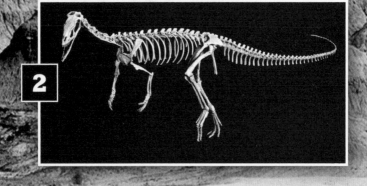

2

to breathing—and which add to evidence that theropod dinosaurs eventually evolved into today's birds.

Hold Me Closer, Tiny Dinosaur

Eodromaeus lived alongside—and now appears to have, in a sense, taken the place of—a very similar dinosaur species, *Eoraptor,* said University of Chicago paleontologist Paul Sereno, also a National Geographic explorer-in-residence. "If you went back 230 million years ago and one of these creatures flitted by, you'd have to wonder which one it was."

> **TRUTH:**
> THE DINOSAUR *EODROMAEUS* WAS ABOUT 4 FEET LONG AND WEIGHED 10 TO 14 POUNDS.

Sereno and his team once thought *Eoraptor* was an ancestor of meat-eating dinosaurs. But due to recent analysis of *Eoraptor* fossils, as well as the discovery of *Eodromaeus,* he now thinks *Eoraptor* was an ancestor of the giant, long-necked, plant-eating dinosaurs called sauropods.

"That's the beauty of dinosaur origins," Sereno said. "Who could predict that these 10- to 15-pound [4.5- to 7-kilogram] creatures—both looking quite similar but eating different things—would end up evolving into things as disparate as *Diplodocus* and *Tyrannosaurus?*"

The reclassification of Eoraptor makes sense, agreed Hans-Dieter Sues, a dinosaur expert at the National Museum of Natural History in Washington, D.C., who was not involved in the study. "One thing that everyone noticed when *Eoraptor* was first discovered was that the back teeth were very odd-looking for a theropod," said Sues, also a contributing editor to the National Geographic News Watch blog. "It had these little leaf-shaped teeth in the back, and those are teeth you don't really find in theropods."

Complete Set

Nearly all of the bones of *Eodromaeus* have been accounted for—considered remarkable for such a small creature. Based on its fossils, scientists think *Eodromaeus,* like its theropod descendants, stood and ran on two legs and had sharp teeth and grasping claws, which the new dinosaur used to snatch the young of other reptiles.

University of Maryland paleontologist Thomas Holtz agreed that *Eodromaeus* is likely an early theropod ancestor. "I think they've got a good case for it here," said Holtz, who wasn't part of the new study. "In terms of characteristics, it does seem to be very, very low in the theropod family tree."

It's not too surprising that *Eodromaeus* and *Eoraptor* looked very similar, he added. Both shared a common ancestor only about ten million years before, which in evolutionary terms is not a very long time. "The closer we get to the common ancestor, the less time they've had to diverge, so they look a lot more like each other," he said. "If you go back far enough, eventually they're the same creature."

Triassic Paradise

The desolate Valley of the Moon in northwestern Argentina, where fossils of *Eodromaeus* and *Eoraptor* were found, was filled with lush forests 230 million years ago, according to study co-author Sereno. "It was a gorgeous environment."

An Avian Blueprint

The *Eodromaeus* had air sacs in its neck bones that might have been used for breathing, which adds to evidence that theropod dinosaurs evolved into today's birds. Modern birds have complex respiratory systems, with two lungs and up to nine air sacs. Such efficient respiratory systems would have boosted meat-eating dinosaurs' metabolisms and enable them to be active and effective hunters.

Eodromaeus and *Eoraptor* shared this Triassic paradise with various other groups of reptiles, including parrot-beaked reptiles that were distantly related to dinosaurs, protomammals, and a number of large crocodile-like creatures. Studying the shared traits between *Eodromaeus* and *Eoraptor* could help scientists paint a picture of the unknown last common ancestor of all dinosaurs, University of Maryland's Holtz said.

That dinosaur Eve, Holtz said, "was probably bipedal, its hands may have already been adapted for grasping . . . and its diet may not have been strictly meat or plants." By contrast, "the first dinosaurs may have been omnivores." ■

Venomous Dinosaur Discovered

A 125-million-year-old dinosaur had more in his arsenal than just sharp teeth. Researchers believe this guy had venom and shocked its prey like a snake.

Jurassic Park (1993) was packed with pseudo-science, but one of its fictions may have accidentally anticipated a dinosaur discovery announced in 2009—venomous raptors.

Finding the Groove

Though a far cry from the movie's venom-spitting *Dilophosaurus,* the 125-million-year-old *Sinornithosaurus* may have attacked like today's rear-fanged snakes, a new study suggests. Rear-fanged snakes don't inject venom. Instead, the toxin flows down a telltale groove in a fang's surface and into the bite wound, inducing a state of shock.

In *Sinornithosaurus* fossils, researchers discovered an intriguing pocket, possibly for a venom gland, connected to the base of a fang by a long groove, which likely housed a venom duct, the study says. *Sinornithosaurus* fangs also feature snakelike grooves in their surfaces.

"The ductwork leading out of the venom gland gave the venom a way to travel to the base of the teeth, where the venom welled up in the grooves," said

> **TRUTH:**
> REAR-FANGED SNAKES DON'T INJECT VENOM— THE TOXIN FLOWS DOWN A GROOVE IN A FANG'S SURFACE AND INTO THE BITE WOUND.

study co-author paleontologist David Burnham of the University of Kansas Natural History Museum and Biodiversity Research Center.

"So when they sank their teeth into tissue of the victim, it allowed the venom, which was really enhanced saliva, to get into the wound."

A Stunning Bite

Turkey-size *Sinornithosaurus,* which likely had feathers, lived in the forests of what's now northeastern China, and was a member of the family Dromaeosauridae. Birdlike *Sinornithosaurus* probably used its longish fangs to put the bite on prehistoric birds, Burnham said. Like rear-fanged snakes and some lizards, the dinosaur probably had nonfatal venom that could shock its victims into a defenseless stupor—allowing *Sinornithosaurus* to eat in peace.

Burnham's research was inspired by the 2000 find of another possibly venomous dinosaur fang and by a recent discovery that today's top lizard predator, the Komodo dragon, has a venomous bite that weakens victims so they can be eaten later.

Though believed to have descended from dinosaurs like *Sinornithosaurus,* today's birds are toothless and so lack a venom delivery system (though some birds do have toxic skin and feathers). But Burnham is more interested in where *Sinornithosaurus'* venom ability came from than how it evolved.

"How primitive is venom really? Does it go all the way back to the archosaurs?" he said, referring to reptiles thought to have predated dinosaurs by 30 million years or more. "These are things people haven't really tested yet." ■

Gotcha! A sculpture of **Sinornithosaurus millenii** *dinosaur on the hunt*

Giant "Roly-Poly"
Rabbit Fossil Found

The king of the bunnies wasn't a sleek, supple creature. Paleontologists believe he was a roly-poly beach bum.

The Easter bunny came early in March 2011 for a few scientists working on the Spanish island of Minorca. The team announced the discovery of Earth's biggest known rabbit species, an oddly unbunny-like giant dubbed *Nuralagus rex*—"the Minorcan king of the hares."

King of the Hares

The 26-pound (12-kilogram) prehistoric species was about six times bigger than the common European rabbit, found on most continents, according to an analysis of several bones. Study leader Josep Quintana is no stranger to giant Minorcan rabbit fossils, though it took a while before he knew exactly how big a find he'd uncovered.

"When I found the first bone I was 19 years old, I was not aware what this bone represented. I thought it was a bone of the giant Minorcan turtle!" said Quintana, a paleontologist at the Institut Català de Palentologia in Barcelona.

TRUTH:

A 53-MILLION-YEAR-OLD RABBIT'S FOOT WAS UNEARTHED IN INDIA.

Odd Body

The animal, which lived about three to five million years ago, had several "odd" features that have never before been seen in rabbits, living or extinct, according to the study. For one, the giant rabbit's "short and stiff" vertebral column meant it couldn't bunny hop. And the relatively small sizes of sense-related areas of its skull suggested that the animal had small eyes and stubby ears—a far cry from modern rabbit ears. "I think that *N. rex* would be a rather clumsy rabbit walking," Quintana said. "Imagine a beaver out of water."

Despite its oddities, *N. rex* has many skull and teeth features found in rabbits—meaning there's "no question" it's a rabbit, according to Brian Kraatz, an expert in rabbit evolution at the Western University of Health Sciences in Pomona, California. "Really, this is a rather typical rabbit head [albeit large] stuck on an atypical rabbit body," said Kraatz, who was not involved in the study.

Living Free and Easy

The newfound rabbit's "roly-poly, tank-like" appearance and weird anatomy may have arisen because of its stress-free lifestyle, Kraatz added. That's because the megarabbit had no predators on Minorca—a luxury that allowed the species to evolve to be bigger and more sedentary, he said. Modern rabbits are small, spry, and have sharp vision to escape predators. "He was probably on an evolutionary vacation," said Kraatz, like an "islander beach bum."

Yet, even though the giant rabbit "didn't have too many cares or worries," Kraatz said, "he got too comfortable, and eventually went extinct." ∎

> **"When I found the first bone I was 19 years old, I was not aware what this bone represented. I thought it was a bone of the giant Minorcan turtle!"**
> **Josep Quintana**
> *paleontologist and study leader, on his initial thoughts upon discovering the* N. rex *fossil*

CHAPTER 11

Under Weird

water
ness

Here there be weirdness! Early maps of the world often showed strange, terrifying monsters swimming in the seven seas, but krakens and sea serpents seem downright boring when compared to what scientists are finding today. One-eyed albino sharks, "walking" fish with hands, giant amoebas, and catfish the size of grizzly bears are living beneath the water's surface, just waiting to surprise someone with their freakishness. So come on in.

Rare "Cyclops" Shark Found

It was a one-eyed, two-finned, swimming, albino people-eater that caught the world's attention in 2011. How common are one-eyed sharks, and what could have caused this creature's strange condition?

Talk about a one-of-a-kind discovery—an extremely rare cyclops shark has been confirmed in Mexico. The 22-inch-long (56-centimeter-long) fetus has a single, functioning eye at the front of its head—the hallmark of a congenital condition called cyclopia, which occurs in several animal species, including humans.

Endangered

Dusky sharks, which average about 10 feet (3 meters) long as adults, are among the slowest growing of the shark species and can live for up to 45 years. Often illegally harvested for its fins, the shark species is rated as "vulnerable" by the International Union for Conservation of Nature.

One-Eyed Monster

In 2011, fisherman Enrique Lucero León legally caught a pregnant dusky shark near Cerralvo Island in the Gulf of California. When León cut open his catch, he found the odd-looking male embryo along with its nine normal siblings. "He said, 'That's incredible—wow,'" said biologist Felipe Galván-Magaña, of the Interdisciplinary Center of Marine Sciences in La Paz, Mexico.

Once Galván-Magaña and colleague Marcela Bejarano-Álvarez heard about the discovery—which was put on Facebook—the team got León's permission to borrow the shark for research. The scientists then x-rayed the fetus and reviewed previous research on cyclopia in other species to confirm that the find is indeed a cyclops shark.

Cyclops sharks have been documented by scientists a few times before, also as embryos, said Jim Gelsleichter, a shark biologist at the University of North Florida in Jacksonville. The fact that none have been caught outside the womb suggests cyclops sharks don't survive long in the wild. Overall, finding such an unusual animal reinforces that scientists still have a lot to learn, Gelsleichter added.

"It's a humbling experience to realize you ain't seen it all yet."

The Eye Has It

The cyclops shark also has other deformities, including albinism, or a total lack of tissue pigments; no nostrils; a bump on its snout; and a spinal abnormality, according to Galván-Magaña. A mother's poor diet—especially a lack of vitamin A—can cause cyclopia in mammals, including people, but it's difficult to determine the cause of the condition in sharks, he said.

> # TRUTH:
> ## ALBINISM IS A RARE CONDITION AMONG SHARKS, OCCURRING MORE OFTEN IN BONY FISHES.

The cyclops baby's eye, located in the middle of its snout, is about 1 inch (2.6 centimeters) wide.

Even so, Galván-Magaña suspects the defects in this case aren't related to pollution. "The fishing area close to Baja California Sur is clear of pollution, is a pristine zone, and we cannot consider it as a cause for abnormalities in dusky sharks," according to the study.

Aside from the handful of deformities, "the rest of the body was apparently normal," with well-formed fins, according to the scientists' study, which has been submitted for publication. Despite its lack of nostrils, the shark's gill slits were normal and well formed, and its mouth contained small teeth, according to the study.

What's in a Name?

Common English-language names for the dusky shark include bay shark, black whaler, bronze whaler, brown dusky shark, brown shark, common whaler, dusky ground shark, and shovelnose.

If it'd been born, the cyclops shark likely wouldn't have lived very long, Galván-Magaña said. The baby's stark white color would've made it more obvious to predators, and its malformed tail would've hindered swimming. There are no previous known cases of albino dusky sharks, "so this report is considered a first record of this malformation for this shark species in Mexico," according to the study.

Fisherman's Prize

After Galván-Magaña's team completed their research, they returned the cyclops shark to León. He regards the cyclops shark as a prize, Galván-Magaña said, and has already preserved his oddity in alcohol. "Lots of people want to buy the shark, but the fisherman says no." ∎

Fish With "Hands"

Found to Be New Species

**Look Ma! They've got hands!
Not much for swimming, the handfish family
uses their fins to walk along the ocean floor.
Here are some of our favorites.**

All of the world's 14 known species of handfish are found only in shallow, coastal waters off southeastern Australia. But about 50 million years ago, the animals likely inhabited regions around the world, as noted by Australia's Commonwealth Scientific and Industrial Research Organisation (CSIRO) scientists. Fossils of the curious creatures have been discovered in the Mediterranean, for example.

Scientists have announced nine newly named species in a recent scientific review of the handfish family. The new-species determinations were made based on a number of factors, including number of vertebrae and fin rays, coloration, the presence of scales and spines, and proportional body measurements, according to review author Daniel Gledhill.

Even among the previously known species, the fish are poorly studied, the review authors add, and little is known about their biology or behavior. There's much more to learn about these freaky fish.

Superfish!

Handfish aren't the only fish with surprising skills. Check out these superfish:

1. **Counting!** Lab experiments have shown that the mosquito-fish, a North and Central American freshwater fish, can count.
2. **Walking!** West African lungfish can both push off a solid surface and move along it using their pelvic fins.
3. **Immortality?** Swimming off the coast of Africa, the coelacanth can live up to 100 years and even longer.

FISH 1 — New, Pink, and Rare

Using its fins to walk, rather than swim, along the ocean floor, the pink handfish is an elusive creature. Only four specimens of the 4-inch (10-centimeter) pink handfish have ever been found, and all of those were collected from areas around the city of Hobart on the Australian island of Tasmania. Though no one has spotted a living pink handfish since 1999, it's taken until now for scientists to formally identify it as a unique species.

The elusive pink handfish is rarely seen in the wild.

FISH 2 — See Spot Walk

The previously known spotted handfish is found on sandy sediments at the bottom of Tasmania's Derwent Estuary and adjoining bays. The fish use their fins to walk along the seabed, where they eat small invertebrates such as worms and crustaceans. Perhaps the best studied species of the handfish family, the spotted handfish is listed as critically endangered by the International Union for Conservation of Nature—meaning it's "facing an extremely high risk of extinction in the wild in the immediate future."

Handfish's slow movements and tendencies to stay within tightly confined habitats would seem to make the fish easy targets for predators. But researchers think handfish have a secret weapon: a toxic skin that kills most

Scientists have studied the spotted handfish the most.

attackers. Anecdotal stories suggest predators may die within an hour of eating a handfish, said CSIRO fish taxonomist Gledhill.

FISH 3 Seeing Red

The red handfish, a previously known species, is listed as vulnerable in Australia, where it's found only around the southern island state of Tasmania. Not much is known about handfish, because their populations are low and they are not often seen in the wild. But researchers

The red handfish lives mostly off the coast of Tasmania.

suggest handfish lay fewer eggs than most other fish species, which means their long-term survival is a concern. Handfish also tend to stay very close to home, so they don't adapt well to new places, said fish taxonomist Gledhill.

FISH 4   Also Available in Purple

Newly described as its own species, the Ziebell's handfish typically has yellow fins, but the species can also appear with a mottled purplish coloration. Ziebell's handfish is found only in small, isolated populations off Tasmania and is listed as vulnerable in Australia. ■

Bright yellow fins typically adorn the Ziebell's handfish.

Wild Fish Uses Tool

Cockle Put to Clever Use

Once thought to be a distinctly human trait, tool use has now been observed in a wide variety of animals, including chimps, dolphins, crows, and octopuses. And now fish are getting in on the act, too.

A blackspot tuskfish off Australia used his noggin to get some dinner. A diver saw the fish take a cockle in its mouth and swim to a nearby rock. Then the fish smashed the cockle again and again against the rock until the shell cracked. Dinner is served, all thanks to clever tool use.

A blackspot tuskfish off Australia carries a cockle to a nearby rock.

Cockles, to Go

A recent study in the journal *Coral Reefs* says the pictures—snapped at a depth of nearly 60 feet (18 meters) in Australia's Great Barrier Reef in 2006—are the first ever taken of a tool-using fish in the wild.

Professional diver Scott Gardner was just about out of air and swimming back to the surface when he heard an odd cracking sound nearby. Swimming over to investigate, he spotted the foot-long (30-centimeter-long) fish at work.

"When Scott showed me his photos, I said 'Wow, this is quite amazing,'" said study co-author Alison Jones, a coral ecologist at Central Queensland University in Rockhampton, Australia.

Most of the previously documented cases of tool use in fish occurred in labs, Jones explained, and it was unknown whether fish used tools in the wild.

Other Tools, Other Users

1. **Crows** can craft twigs, leaves, and even their own feathers into tools.
2. **Orangutans** make whistles from bundles of leaves that are used to help ward off predators.
3. **Elephants** have been seen dropping logs or rocks on electric fences to short-circuit them.
4. **Dolphins** may carry marine sponges in their beaks to stir ocean-bottom sand and uncover prey.
5. **Sea otters** use stones to hammer abalone shells off the rocks and crack the hard shells open.
6. **Gorillas** have been known to test water depth with large branches.
7. **Macaques** living near a Buddhist shrine in Thailand are known to pull out visitors' hair to use as floss.
8. **Degus,** small rodents related to chinchillas, were taught how to use rakes to find food.

Tooling Around, or Not

Not everyone agrees the shellfish-smashing fish is actually using a tool. "It's taken six years to publish this research because reviewers argued back and forth about whether it was a true example of tool use or not," co-author Jones said.

At issue is whether the tuskfish behavior fits the classic definition of tool use, which requires an animal to actually hold or carry the tool and use it to manipulate another object.

But that definition was based on human and primate behavior, said Jones and her team, and can't be applied to fish.

"We argue that tuskfish have a different type of intelligence that's evolved for their specific environment . . . They don't have hands, and they have resistance in the water, so you can't expect them to use a tool the way a primate would."

Fangs for Dinner

If scientists accept that tuskfish are indeed capable tool users, then it raises all sorts of interesting questions about fish behavior, the team says. For example, do the tuskfish use the same spot each time to crack open shellfish? If they do, it "adds weight to the long-held view that fish have a cognitive map and they can remember where to go," Jones said.

It may be a while before the team has an answer. The behavior hasn't been seen since 2006, but the researchers hope to catch a tuskfish in the act again soon. "We're hoping to set up a camera at one of these sites at some point and get more information, but it's technically demanding," Jones said.

If tuskfish can use tools in the wild, then it's possible other fish species can too—and that could have broader implications for fisheries management, Jones said. "If older fish are teaching younger fish these patterns of behavior, then by taking out all of the older"—and therefore bigger and more attractive to fishers—"fish, will the younger fish be less able to forage effectively?" Jones said. "That's one of the critical questions that this observation raises." ∎

> **TRUTH:** WHEN A DOMINANT MALE TUSKFISH DIES, THE LARGEST FEMALE OF THE GROUP CHANGES SEX TO TAKE ITS PLACE.

Piranhas Bark

Three Fierce Vocalizations Deciphered

If we could talk to the animals, the perusings of piranhas might be a bit one-note. Underwater microphones have picked up on their limited— and uniformly grumpy—"vocabulary." So what are they saying?

Piranhas, it turns out, can be excellent communicators, a new study suggests. But don't get the idea they're going soft—their barks, croaks, and clicks likely mean "Leave me alone," "I might bite you," or "Now I'm really angry!"

> "Eventually, if we understand the behavior that's associated with the sounds, we might be able to listen to the sea and explain to fishermen: 'Now's not the best time to start fishing.'"
>
> **Eric Parmentier**
> *lead researcher,
> University of Liège, Belgium,
> on deciphering piranha sounds*

Fish Croaks

Researchers knew picking up red-bellied piranhas—among the few types dangerous to humans—could prompt croaks from the fish. Even so, no one had studied their sounds in water or provided good evidence for the barks' evolutionary role.

Now a fish tank, an underwater microphone, and a video camera have helped uncover three different piranha calls—all tied to a variety of grumpy behaviors. "We knew piranhas were able to make sounds but were not satisfied with the explanation for how they do it," said biologist Eric Parmentier of the Université de Liège in Belgium. "We wanted to know how they do this and what these sounds might mean to other fish."

Grunts and Groans

Twenty-five species of piranhas exist in the wild today, but only "two or three" species pose a threat to humans, Parmentier said. In particular, the red-bellied piranha's voracious appetite for fresh meat is a big reason many scientists have shied away from studying any in-water vocalizations, he added.

Parmentier and study co-author Sandie Millot of the University of Algarve in Portugal, though, used their tech-heavy technique to link three distinct sounds to three aggressive piranha behaviors. A repetitive grunt was tied to a visual face-off, as if to say, "Get away from me."

A second call resembling a low thud was associated with circling and fighting. Both of these calls, the researchers discovered, were made using a fast-twitching muscle that runs along a piranha's swim bladder—an air-filled organ that helps fish maintain their buoyancy.

If fellow piranhas didn't heed these warning calls, the aggressor would begin chasing the neighboring fish and making a third type of sound by faintly gnashing its teeth.

TRUTH: PIRANHAS ARE OMNIVORES AND WILL EAT ALMOST ANYTHING, INCLUDING FISH, SNAILS, INSECTS, AQUATIC PLANTS, AND EVEN LARGER MAMMALS AND BIRDS.

Language Lessons

In the future, Parmentier and Millot would like to go to South America to record the red-bellied piranhas—and the other couple of dozen piranha species—in their native environment. "The nature of these fishes are quite special, and I suspect they can make more than three sounds," Parmentier said, adding that they may also use them for hunting or mating.

"Also, there are only recordings for a few species of piranhas. We'd like to see what other species are capable of." ∎

Squid Males Evolved

Shot-in-the-Dark Mating Strategy

Mating with anything with eight arms—male or female—pays off in the dim depths of the ocean, study says.

When it comes to mating, some male squid aren't very picky: They copulate just as often with other males as with females, a new study says.

That's because would-be suitors of the hand-size species *Octopoteuthis deletron,* which live in the murky depths of the eastern Pacific Ocean, can't easily tell the males from the females, the research shows.

Take a Chance on Me

"They can see each other, but they are not able very well to distinguish between the sexes at the distance at which they decide, 'I'm going to mate' or 'I'm not going to mate,'" said study leader Hendrik Hoving, of the Monterey Bay Aquarium Research Institute in California. So "males mate with basically any member of the same species . . . They just take a chance."

It's also hard to tell he from she: A female squid's defining feature is a patch of wrinkled skin. The result is a strategy that the study authors call "a shot in the dark"—it's just not worth it to males to make sure their partner is the right gender.

> **TRUTH:**
> **MOST SQUID HAVE THREE HEARTS.**

Slap Happy

For the study, Hoving and colleagues recorded squid via robotic submarines in the dark—1,300 to 2,600 feet (400 to 800 meters) underwater. The scientists observed more than a hundred male and female squid, and they found that just as many male squids as female squids bore sperm packets on their bodies—showing that males slap a sperm packet on just about anything with eight arms.

When he finds a suitable partner, the male uses his large penis to transfer multiple sperm packets to the male or female. These break open into smaller sperm sacs that attach to his partner's mantle, fins, and arms. But the "love affair" ends there: The squid, which lead a solitary existence, die shortly after mating.

Same Sex Mating Is Rare

Nathan Bailey, of the U.K.'s University of St. Andrews, said the study team "makes a pretty good case" for their claims about the male squid's lack of choosiness. Very few species show such high levels of what biologists call same-sex sexual behavior, Bailey, who wasn't involved in the research, said by email. "Some primates or dolphins do, but this study puts *O. deletron* on the higher end of the scale."

Homosexual Animals

There are a number of well-documented instances of homosexual or bisexual behavior among animals:

1. Roy and Silo are two-male chinstrap penguins at New York's Central Park Zoo who were together for six years, from 1999 until 2005.
2. Some male ostriches court only their own gender, and there are pairs of males flamingos that mate, build nests, and even raise foster chicks.
3. Japanese female macaques were recently discovered engaging in intimate acts.
4. Studies suggest that 75 percent of bonobo sex is nonreproductive and that nearly all bonobos are bisexual.
5. Other animals appear to go through a homosexual phase before they fully mature—male dolphin calves often form temporary sexual partnerships.

Hoving acknowledged that his research results can become fodder for jokes. "But I don't really care," he said. "I'm interested in deep-sea animals and how they're capable of living in that environment, and one of the challenges is finding the opposite sex." ■

Great White Shark Jumps on Boat, Stressing Everyone

Most fish stories sound too good to be true, but this one sounds just plain scary. You decide.

Marine researchers chumming the ocean to lure sharks closer to their vessel off South Africa's southwestern Cape coast got more than they bargained for when a half-ton (500-kilogram) great white shark leaped into their boat. It happened in July 2011, near Seal Island, off Mossel Bay, in a part of the ocean famous for its "flying" sharks.

History Lesson

According to the International Shark Attack File, which has attack data going back to 1580, the deadliest sharks to humans are:

1. Great white shark : 431 attacks
2. Tiger shark: 169 attacks
3. Bull shark: 139 attacks

"We go out into the bay everyday to get data on the shark population," said Enrico Gennari, director of Oceans Research, an independent research organization that works with universities and runs public awareness programs to teach people about sharks. Oceans Research has also collaborated with National Geographic to produce television documentaries about sharks and other predators.

"Our team was chumming to attract sharks to the boat so that we could photograph their fins, which, like human fingerprints, are a way to identify individual animals," Gennari said. "They waited for four or five minutes, but nothing happened until there was an enormous splash and a shark landed in the boat."

Off the coast of South Africa, great white sharks are known for their leaping abilities.

The team's first thought was to make sure everyone was well clear of the giant predator, which was thrashing around on the deck. "They moved everyone out of the way, then radioed us," Gennari said. He and a colleague arrived at the boat within minutes. It was not possible to push the flailing animal back into the sea by hand, and an attempt to pull it off the vessel with a rope attached to the second boat also failed.

Shark Overboard

"We radioed the harbor and said we were on the way, and that we needed a crane. It took about 20 minutes, and the whole time we were splashing water on the shark's gills, trying to help it breathe," Gennari said. "At the harbor, a pipe was placed into the shark's mouth to pump water over its gills to keep it alive. The crane was used to hoist the animal by its tail, a risky maneuver because its great weight out of the water could have damaged its spine and internal organs. But the only other option we had was to let it die on the boat."

The shark was lowered into the water, but it stranded itself on a harbor beach. Attempts to push it back into the water by hand failed, so the Oceans Research team lashed the animal to the side of a boat and drove it out to sea. After half an hour of the assisted swim through the ocean the shark seemed to recover, slapping its tail strongly, Gennari said. The shark was released and it swam away, he added.

> **TRUTH:**
> **THE WORLD'S LARGEST SHARKS ARE TYPICALLY FEMALE, NOT MALE.**

"We don't know whether it's still alive; we hope to see it again soon," Gennari said. The specimen is well known to the Oceans Research team because it had been photographed regularly, and it is readily identified by its unique dorsal fin.

"This was the first, and hopefully last, time a shark has jumped into our boat," Gennari said. "It was quite stressful for everyone, both for the shark and the humans. But the people were safe and apparently the shark survived, so it really could not have ended better."

White Sharks Can Jump

This is not the first time a shark has jumped into a boat in this part of South Africa's coastal waters. "There have been several incidents of leaping great white sharks in the past," Chris Fallows of Apex Shark Expeditions told National Geographic. Fallows has been monitoring and photographing breaching great white sharks for 20 years. "In 1976 in False Bay, it happened on two different occasions," Fallows wrote. "One of the incidents resulted in a very badly injured fisherman. Sadly in both cases, the shark died."

Apex Shark Expeditions records around 600 to 700 predatory events every year at Seal Island. "Often during these events we see spectacular breaches. It is thus inevitable that over the years we have had a few sharks jump close to the boat, but due to a huge emphasis being placed on safety, good fortune, as well as a fairly large boat, we have thankfully not had a shark land in [our] boat."

So what could cause a shark to jump on a boat? "There can be several things," Fallows explained. "In most cases the sharks breach while in pursuit of something, be it a seal, or fish, or a decoy. Occasionally however, great whites and some other species perform what is known as a natural breach. This type of breach takes place for no apparent reason, although it is speculated that it could have some form of social function of communication or dominance. These breaches are often very high, the mouth of the shark is closed, and it is often when there are several sharks in the immediate vicinity." ■

Big Boy

In fall 2009, researchers landed the largest shark ever— a two-ton-plus, 17.9-foot-long male. After fitting the fish with a satellite-tracking tag and taking a blood sample, they released him back into the waters off Mexico's Guadalupe Island.

Giant "Amoebas"

Found in Deepest Place on Earth

When scientists venture to Earth's most extreme places, they are often surprised by the life-forms they find. This time, they traveled to the deepest place on Earth and found a giant one-celled creature.

Huge "amoebas" have been spotted in the Mariana Trench, the deepest part of the world's oceans.

The giants of the deep are so-called xenophyophores, sponge-like animals that—like amoebas—are made of just one cell. They were found during a July 2011 research expedition run by the Scripps Institution of Oceanography in La Jolla, California.

TRUTH: XENOPHYOPHORES CAN CONCENTRATE HIGH LEVELS OF LEAD, URANIUM, AND MERCURY AND ARE THUS LIKELY RESISTANT TO LARGE DOSES OF HEAVY METALS.

Attack of the 4-inch Amoeba

The animals are about 4 inches (10 centimeters) long—among the largest single-celled organisms known to exist. The creatures were discovered at depths of 6.6 miles (10.6 kilometers). That breaks a previous record for xenophyophores found in the New Hebrides Trench at 4.7 miles (7.6 kilometers).

Xenophyophores represent "one of the few groups of organisms found exclusively in the deep sea," said Lisa

Levin, a Scripps oceanographer who studied the expedition's data. "If any creatures should be able to live at the ocean's greatest depth, then xenophyophores certainly should be among them."

Lower the Dropcams

The Mariana xenophyophores were seen in footage from Dropcams, free-falling devices equipped with lights and digital video that were developed by the National Geographic Society. Protected by thick walls of pressure-resistant glass, the Dropcams were baited to attract whatever marine life might be lurking in the deep. Expedition scientists also saw, for instance, the deepest-swimming jellyfish to date.

"The deep sea is the largest biome on Earth and holds much of the diversity on the planet—[yet it's still] largely undescribed," Levin said in an email to National Geographic News.

According to Jon Copley, a marine biologist at the United Kingdom's University of Southampton, "many of the major discoveries in deep-sea biology have come from making direct observations at the seafloor."

"The Dropcam is a great tool for the future because it can help us see more of what's down there for less cost than using ROVs or submersibles," he said via email. For instance, "finding xenophyophores far deeper than before shows how much we still have to learn about our ocean's depths and their inhabitants."

Tullis Onstott, an expert in deep-sea microorganisms at Princeton University, also called the xenophyophore discovery "fantastic." "Who knows what's next, behemoth nematodes?" he said by email. ∎

> "The xenophyophores are . . . fascinating giants that are highly adapted to extreme conditions but at the same time are very fragile . . . These and many other structurally important organisms in the deep sea need our stewardship as human activities move to deeper waters."
>
> **Lisa Levin**
> *deep-sea biologist, Scripps Institution of Oceanography*

Sawfish Snout Has Sixth Sense

Splits Prey in Half

The sawfish has a nose like a Swiss Army knife—it can slice up prey, probe the bottom for food, and serve as an "antenna" to detect electric fields of prey.

They may not see dead people, but sawfish use a sixth sense based in their long snouts to hunt and dismember prey, new research shows for the first time.

Previously, scientists had suspected that sawfish—large ocean and freshwater fish found throughout the tropics—use the saw, a cartilaginous extension of the skull, to probe sand or mud for prey. The saw also serves as a weapon, the new research suggests. Lateral swipes, observed during experiments in the lab, can split smaller fish in half.

Ovo-What?

Sawfish are ovoviviparous, meaning their offspring are in eggs, but the eggs develop inside the mother's body. The young are nourished by a yolk sac. Depending on the species, gestation may last from several months to a year. Sawfish pups are born with their saw fully developed, but it is sheathed and flexible to avoid injuring the mother at birth.

Invisible Touch

Now, preliminary experiments also suggest that the fish's long, tooth-lined saw is full of pores that can detect movements or electric fields of passing prey—acting as a sort of "distant touch," Barbara Wueringer, a sensory neurobiologist at the University of Queensland in Australia, said by email. This skill is especially handy for nosing out dinner in murky or dark waters, Wueringer said.

"We know so little about sawfish, even though these animals can grow really big"—up to 16 feet (5 meters). "To know that the saw acts like an antenna that can sense prey is amazing."

Making Sense of Sawfish

In addition to observing the animals, Wueringer dissected several sawfish that had been accidentally caught or died naturally. She found that the sawfish's saws were full of tiny pores that signal the ability of an animal to detect the electric fields present in all living animals. Sharks and rays have these pores, as do some fish such as lungfish and even some mammals such as the echidna.

By making "maps" of the skin of four species of rare sawfish, Wueringer pieced together where the pores are distributed on the saws. She then compared this data with pore placement on two species of shovelnose rays.

> **TRUTH:** THE "TEETH" ON THE SAWFISH'S SNOUT ARE NOT TRUE TEETH, BUT MODIFIED SCALES. ITS REAL TEETH ARE LOCATED INSIDE ITS MOUTH, WHICH IS ON THE FISH'S UNDERSIDE.

This fish's saws are lined with pores that detect electric fields.

Determining where pores are most concentrated gives clues about animals' feeding behaviors, she said. "For example, rays have their eyes on the upper side of their head, but the mouth is on the lower side. Pores that can detect electric fields around the mouth allow these animals to sense a fish when they are trying to ingest it—but cannot see it," she said.

In the sawfish, the pores were most concentrated on the upper sides of their saws, which should enable the predators to stalk fish in the space above their saws.

Research May Help Rare Species

Overall, Wueringer hopes the research will help conservationists learn more about sawfish, especially the four species she studied, whose last stronghold is a remote bay of northern Australia.

In general, sawfish have dwindled dramatically in recent years, largely due to overfishing, both intentional and accidental—the sawfish's serrated snouts make the animals especially vulnerable to entanglement in fishing nets, according to the International Union for Conservation of Nature.

"In order to protect endangered species, we need to know as much as we can about them," she said. "How a species catches its prey, and also which senses are involved in detecting the prey, is part of the basic understanding of a species." ■

"Bizarre" Octopuses

Carry Coconuts as Instant Shelters

Coming soon to an ocean near you—the amazing disappearing octopuses! A new study shows that these smart creatures are using coconut shells as portable places to hide.

Octopuses have been discovered tiptoeing with coconut-shell halves suctioned to their undersides, then reassembling the halves and disappearing inside for protection or deception, a new study says.

Peek-a-boo! A veined octopus hides in coconut shells.

"We were blown away," said biologist Mark Norman of discovering the octopus behavior off Indonesia. "It was hard not to laugh underwater and flood your [scuba] mask."

The coconut-carrying behavior makes the veined octopus the newest member of the elite club of tool-using animals—and the first member without a backbone, researchers say.

Coconuts to Go

A team led by biologist Julian Finn of Museum Victoria in Melbourne, Australia, was observing 20 veined octopuses (*Amphioctopus marginatus*) on a regular basis. The researchers noticed that the animals were frequently using their approximately 6-inch-long (15-centimeter-long) tentacles to carry coconut shells bigger than their roughly 3-inch-wide (8-centimeter-wide) bodies.

An octopus would dig up the two halves of a coconut shell, then use them as protective shielding when stopping in exposed areas or when resting in sediment. This action, on its own, astonished the team. Then they noticed that the octopuses, after using the coconut shells, would arrange them neatly below the centers of their bodies and "walk" around with the shells—awkwardly.

TRUTH: AN OCTOPUS CAN HAVE NEARLY 2,000 SUCKERS ON ITS ARMS.

"I've always been impressed by what octopuses can do, but this was bizarre," said study co-author Norman, senior curator for mollusks at Museum Victoria.

To carry the shells, a veined octopus has to stick its arms out and over the edges of the coconut and walk around as if on stilts—making the octopus, while in motion, more vulnerable to predators—study leader Finn explained.

"An octopus without shells can swim away much faster by jet propulsion," he said. "But on endless mud seafloor, where are you fleeing to?" In other words, a coconut-carrying octopus may be slow, but it's always got somewhere to hide.

What a Tool

So, what makes the veined octopus's behavior tool use, versus, say, the hermit crab's use of seashells as armor?

Worn nearly constantly, a hermit crab's adopted shell isn't considered a tool, because it's always useful. Tools, by definition, provide no benefit until they're used for a very specific purpose—showing that the animal is capable of what you might call advance planning. The octopus's coconut carrying qualifies as tool use, Finn said, because the shells provide only "delayed benefits."

Octopuses of many species are well known for their intelligence. In captivity they've been known to navigate mazes, seem to be able to remember past events, and are cunning escape artists.

Octopus Tool Use Surprising

Tool use, once thought to be a uniquely human behavior, is seen as a sign of considerable mental sophistication among nonhuman animals. It's been known for years now that chimpanzees use whole "tool kits," that some dolphins attach sponges to their beaks for fishing, and that crows fish for insects using sticks and leaves, for example.

> "While the octopus carries the coconut around there is no use to it—no more use than an umbrella is to you when you have it folded up and you are carrying it about . . . The coconut becomes useful to this octopus when it stops and turns it the other way up and climbs inside it."
> **Tom Tregenza**
> *evolutionary ecologist*

Even so, the octopus discovery stands apart. "I really wasn't expecting to see tool use appear in cephalopods"—squid, cuttlefish, and octopuses—said biological anthropologist Craig Stanford, co-director of the Jane Goodall Research Center in Los Angeles, who wasn't involved in the new study.

That the octopuses weren't using their tools to rustle up dinner only added to Stanford's surprise. "Even chimps," he added, "do not use natural materials to create shelters over their heads." ■

Lobster Caught

"Half Cooked" in Maine

Batman fans will remember Two-Face, the villain with a mug that's half handsome and half gruesome. In 2006, a Maine lobsterman caught a different kind of two-faced prey—a lobster that looks half raw and half cooked.

Alan Robinson of Steuben, Maine, hauled up an unbelievable catch near the town of Bar Harbor in 2006: It was a rare two-toned lobster. Half of the animal was mottled brown while the other was bright orange—the color lobsters turn after they've been boiled. In his 20 years of catching the crustaceans, Robinson says, he has never seen anything like it.

Robinson spotted the animal while bringing in his catch. "I thought someone was playing a trick on me," he told the *Bangor Daily News.* "Once I saw what it was . . . it was worth seeing."

TRUTH: LOBSTERS HAVE POOR EYESIGHT BUT HIGHLY DEVELOPED SENSES OF TASTE AND SMELL.

Two-Toned Lobster

He wanted others to see it, too, so Robinson donated his unusual catch to Maine's Mount Desert Oceanarium, where experts were able to shed some light on the find. Two-toned lobsters, they explain, are rare but not unheard of. The shells of American (also called Maine) lobsters usually sport a combination of yellow, red, and blue pigments. But the animals grow symmetrically, with each half of the

The rare half-brown, half-orange lobster

body developing independently of the other. In the case of Robinson's catch, half of the lobster's shell was lacking the blue pigment, giving its body the appearance of having been cooked to a turn.

All this makes Robinson's 50-50 find one for the record books, the Oceanarium's staffers say. The aquarium has received only three two-toned lobsters in 35 years. Lobsters can sport other odd colorations—1 in 2 million is all blue and 1 in 30 million is all yellow. But the odds of finding one that's exactly half blue and half orange are much higher: 1 in 50 million. ■

Record-Setting Catch

The largest lobster recorded was caught off the coast of Nova Scotia, Canada. It weighed 44.4 pounds and was between 3 and 4 feet long. Scientists think it was at least 100 years old.

Weird Fish With Transparent Head

Bringing new meaning to the phrase "I see right through you," the Pacific barreleye lets you see exactly what's on—and in—its mind.

With a head like a fighter-plane cockpit, a Pacific barreleye fish shows off its highly sensitive, barrel-like eyes—topped by green, orblike lenses. The fish, discovered alive in the deep water off California's central coast by the Monterey Bay Aquarium Research Institute (MBARI), is the first specimen of its kind to be found with its soft transparent dome intact. The

A see-through Pacific barreleye fish

6-inch (15-centimeter) barreleye *(Macropinna microstoma)* has been known since 1939—but only from mangled specimens dragged to the surface by nets.

I'm Looking Through You

The beady bits on the front of the Pacific barreleye fish aren't eyes but smell organs. The grayish, barrel-like eyes are beneath the green domes, which may filter light. In this picture the eyes are pointing upward—the better to see prey above in the darkness of the barreleye's deep-sea home.

Since the eyes are upright tubes, "It just looked like [they only] looked straight up," MBARI marine technician Kim Reisenbichler said. But by watching live fish, the scientists discovered that the eyes can pivot, like a birdwatcher pointing binoculars.

The transparent-headed Pacific barreleyes may steal fish from siphonophores, jellyfish that can grow to more than 33 feet (10 meters) long. The barreleye's flat, horizontal fins may allow it to swim very precisely among the siphonophore's stinging tentacles—and if the fish fumbles, the clear, helmet-like shield may protect its eyes.

> "Look. I'm all for sushi. But see-through-shi?"
> **Stephen Colbert**
> comedian, on the discovery of the barreleye fish

Deep Down

The barreleye lives more than 2,000 feet (600 meters) beneath the ocean's surface, where the water is almost inky. The transparent-headed fish spends much of its time motionless, eyes upward. The green lens atop each of the fish's eyes filters out what little sunlight makes it down from the surface. Then the eyes rotate forward to follow the prey, allowing the fish to home in on its meal. ■

Strange Sea Species
Found off Greenland

Five of the weirdest-looking fish you've ever seen are now swimming in the waters off Greenland.

A recent study found 38 fish species swimming in Greenland's waters for the first time. Ten of the species new to Greenland are new to science too. Despite their outwardly weird appearances, there is much to be learned from these fish.

Led by biologist Peter Møller of the Natural History Museum of Denmark in Copenhagen, the study says that rising ocean temperatures due to global warming—which could be drawing unfamiliar fish to the region—and increased deep-sea fishing may account for the spike in fresh fish faces seen off Greenland, according to the study published in the journal *Zootaxa*.

FISH 1 Iceland Catshark

The Iceland catshark species, including the fish caught

during the study period, is among several sharks recently found in Greenland waters for the first time. The small shark has been found in other oceans at depths of between 2,645 to 4,625 feet (800 to 1,410 meters), where it feeds on fish, marine worms, and crustaceans such as lobster and crabs.

FISH 2 — Longhead Dreamer

Looking like a creature from the *Alien* movies, this not-so-beautiful "longhead dreamer" anglerfish *(Chaenophryne longiceps)* was until recently an alien species to Greenland waters. The nightmarish-looking dreamer grows to a not-so-monstrous 6.7 inches (17 centimeters) in length.

FISH 3 — Monkfish

It may be unappetizing to look at, but this newly arrived species of anglerfish, *Lophius piscatorius*—that's "monkfish" to seafood fans—could prove a tasty addition to Greenland's fishery, according to study leader Møller.

Though monkfish remain rare in Greenland, they appear to be taking advantage of the island's warmer sea temperatures—as are fellow relatively

shallow-water species, including Mueller's pearlsides, whiting, blackbelly rosefish, and snake pipefish. "Monkfish is so expensive and popular" that it stands out as a potential commercial species from all the other new fish recorded in the survey, Møller said.

FISH 4 | Atlantic Football Fish

Scaly oddities trawled up from seas around Greenland since 1992 include the Atlantic football fish, a type of anglerfish that lures prey by waggling its fleshy "bait."

The stubby, deep-sea species belongs to an anglerfish group in which the males attach themselves to the much larger females like parasites. The tiny male—little more than a sperm donor—is nourished by the female until her eggs are fertilized.

FISH 5 | The Swallower

Chiasmodon harteli belongs to a group of fishes known as swallowers because of their ability to swallow prey larger than themselves. It's also among the 38 species never before seen off Greenland. Hundreds of

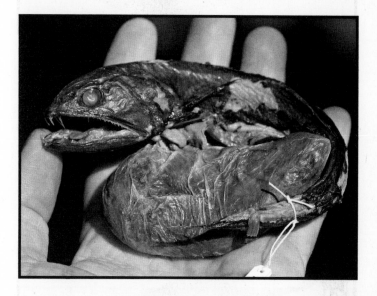

yards above *Chiasmodon harteli*'s deep habitat, Greenland has been extensively fished for more than a century.

At these shallower depths, it's reasonable to assume that "any unknown species of fish occurring in today's catches are in fact new in the area," the study team writes. ■

Grizzly Bear-Size Catfish

Caught in Thailand

Measuring as big as a grizzly bear, a huge catfish caught in 2005 in northern Thailand still holds the record for the largest freshwater fish ever.

It wasn't your typical fish story when a team of fishermen on the Mekong River struggled for more than an hour to haul the creature in. It tipped the scales at 646 pounds (293 kilograms) and measured nearly 9 feet long (2.7 meters). Thai fisheries officials had hoped to release the adult female after stripping it of eggs for a captive-breeding program. Unfortunately, the fish didn't survive its ordeal.

It died and was later eaten by villagers.

This giant catfish measured nearly 9 feet (2.7 meters) long.

World's Biggest Catfish?

The giant catfish has been the focus of a World Wildlife Fund (WWF) and National Geographic Society project to study the planet's biggest freshwater fish. "It's amazing to think that giants like this still swim in some of the world's rivers," said project leader Zeb Hogan, a National Geographic Society emerging explorer and a WWF conservation science fellow. "We believe this catfish is the current record-holder," Hogan added. "I have heard of 3-meter-plus (10-foot) catfish in Bulgaria, 500-kilogram (1,100-pound) stingrays in Southeast Asia, and 5-meter (16-foot) arapaima in the Amazon, but up until now we have not been able to confirm these reports."

Other contenders for the title of world's largest freshwater fish include the Chinese paddlefish and dog-eating catfish—another Mekong giant. Hogan says such big species are poorly studied and in urgent need of protection. "In many locations they are now so rare that the opportunity for documentation and study may soon be lost," he said.

> "Fishing is the most easily identifiable threat to the Mekong giant catfish. Dams, navigation projects, and habitat destruction also threaten the giant catfish. In the Mun River, the largest tributary to the Mekong, a dam blocks the migrations of giant catfish."
>
> **Zeb Hogan**
> *researcher and conservationist, Mekong Fish Conservation Project*

Meditating Megafish

International efforts are underway to save the species. It is now illegal in Thailand, Laos, and Cambodia to harvest the giant catfish. However, enforcement of fishing restrictions in many isolated villages along the Mekong is nearly impossible, and illicit and bycatch takings do continue.

There's a long tradition of giant catfish fishing in Thailand and Laos. Hogan says cave paintings of the fish in northeast Thailand show it has captured the imagination of people living along the Mekong for more than a thousand years. "Mekong people believe it's a sacred fish because it persists on plant matter and 'meditates' [in the deep, stony pools of the Mekong River]—somewhat like a Buddhist monk," Hogan said. ∎

New Jellyfish

Attacks Other Jellies

Off the Florida Keys swims a newly discovered species of jellyfish nicknamed the "pink meanie." It has hundreds of stinging tentacles used to hunt its favorite prey: other jellies.

When pink meanies were first observed in large numbers in the Gulf of Mexico in 2000, they were thought to be *Drymonema dalmatinum,* a species known since the late 1800s and usually found in the Mediterranean Sea, the Caribbean Sea, and off the Atlantic coast of South America.

Recently, though, scientists using genetic techniques and visual examinations have revealed that this pink meanie is an entirely new species—*Drymonema larsoni,* named after scientist Ron Larson, who did some of the first work on the species in the Caribbean.

> # TRUTH:
> ## A JELLYFISH CAN BE AS SMALL AS A THIMBLE OR AS LARGE AS TWO WASHING MACHINES.

A New Family

Moreover, the pink meanie appears to be so different from other known scyphozoans, or "true jellyfish," that it forced the scientists to create a whole new animal family, a biological designation two levels above species. The new scyphozoan family—the first since 1921—is called Drymonematidae and includes all *Drymonema* species.

"They're just off by themselves," said Keith Bayha, a marine biologist at the Dauphin Island Sea Lab in Alabama. "As we started to really examine

Hundreds of stinging tentacles dangle from a "pink meanie"—a new species of jellyfish with a taste for other jellies.

Drymonema both genetically and morphologically, it quickly became clear that they're not like other jellyfish and are in their own family."

Bayha and Michael Dawson, an expert on the evolutionary history of marine creatures at the University of California, Merced, detailed the new *Drymonema* jellyfish species and family in the journal *The Biological Bulletin*.

A Taste for Its Own Kind

According to the new analyses, this Gulf of Mexico *Drymonema* species is genetically distinct from its Mediterranean cousin, *D. dalmatinum*. Regardless of where they live, all *Drymonema* species have an appetite for moon jellyfish, which the *Drymonema* feed on almost exclusively as adults. Larger *Drymonema* can ensnare multiple moon jellyfish at once—one had been found with 34 moon jellyfish in its tentacles.

"They just spread their tentacles out, and as soon as they come into contact with a moon jellyfish, they get more tentacles around them and pull them in," the Dauphin Island Sea Lab's Bayha explained.

Adult *Drymonema* do the majority of their digestion using specialized "oral arms" that dangle alongside their tentacles. The oral arms exude digestive juices, which break down the prey.

> "As a rule, jellyfish tend to be relatively understudied compared to other animals, and we are constantly uncovering new information fundamental to our understanding of these interesting animals and how they interact with humans and the marine environment."
> **Keith Bayha, Ph.D.**
> biologist, Dauphin Island Sea Lab

Gulf War

Drymonema can vary greatly in size. Some are only a few inches across, while others can grow to several feet in diameter. "They just keep growing, but most jellyfish live only a year," Bayha explained. "They'll breed and then they'll kind of stop eating and, in essence, shrivel up, and die."

While *Drymonema* jellyfish feed mainly on other jellyfish, the stinging cells in their tentacles are potent enough to be felt by humans. "They're really bad stingers," Bayha said. "The more tentacles come into contact with you, the worse the sting is going to be. And these guys have hundreds and hundreds of tentacles."

Since many jellyfish look very similar, past researchers assumed that there are very few jellyfish species. But UC Merced's Dawson has revealed many cryptic jellyfish—jellies that look the same but are actually separate species.

While the discovery that a single global species might actually be multiple species may seem trivial, it can become important when studying jellyfish ecology, since different species might behave differently. "It changes the way in which we can study these guys and how they interact with humans and the marine environment," Bayha said. "And they're being recognized more and more as a major pest around the world." ■

TRUTH:
A GROUP OF JELLYFISH IS CALLED A "SMACK."

Small Squid Have Bigger Sperm

Chalk one up for the little guy: Smaller squid males—called "sneakers"—have evolved larger sperm, a perfect match for a female reservoir reserved just for them.

Scrawny spear squid have size where it matters—in their sperm. A new study shows that the smaller males—called sneaker squid—pack heftier sperm than big spear squid. Also, the sneakers' sperm are tailored for a completely separate sperm reservoir, near the female's mouth, researchers found.

Large Tactics: Light Show

Female spear squid mate willingly only with big males, which court the females with natural, multicolored, pulsating lights—bioluminescent mating displays that the smaller males don't exhibit.

Once the female selects a male, he holds her above him and inserts his arm, holding a sperm packet, into her oviduct, a passage through which his sperm

> **TRUTH:**
> THE SPEAR SQUID IS THE FIRST SPECIES KNOWN TO HAVE INDIVIDUALS THAT PRODUCE TWO SEPARATE TYPES OF SPERM.

travel to her ovaries. He'll then watch over her until she spawns to try to ensure no other males fertilize her eggs.

Small Tactics: Stealthy & Quick

Yet just as the female begins to lay her eggs on the seabed, a smaller squid, a sneaker, may dart over and mate with her head-to-head.

A female is typically "willing to accept sneaky matings, as she has a specific sperm storage organ near [her] mouth for sneaker males," study first author Yoko Iwata, a postdoctoral research fellow at the University of Tokyo, said by email.

Though the larger males end up fertilizing most of the eggs, the sneaker males' behavior ensures that the smaller squid still have a chance at fathering some of the offspring.

Bigger Sperm Matters

For the experiment, Iwata and colleagues dissected commercially fished squid, collecting sperm from the males and from the females' two sperm storage sites.

The team then measured the sperm cells under the microscope and discovered that the larger males' sperm—deposited inside the female—were an average of 73 millionths of a meter long. The sneaker male sperm averaged about 99 millionths of a meter. The scientists also collected eggs from live female squid and artificially inseminated the eggs, confirming that both sperm types were fertile.

> "Sperm size is likely to be an adaptation to fertilization environment, either inside the female or externally, rather than competition between sperm, because the fertility and motility of sneaker and consort sperm were the same."
>
> **Yoko Iwata**
> *postdoctoral research fellow, University of Tokyo*

What's more, the two types of sperm swam at the same speed, suggesting that the bigger sperm don't have an edge in competition with the smaller sperm. Rather, the sneakers' bigger sperm likely evolved to suit the females' sperm reservoir and the surrounding seawater.

In other words, each of the female's two sperm receptacles, and waters around these receptacles, may have varying characteristics—such as pH, salinity, or concentration of gases and nutrients—that favor one sperm size over another.

The results are the first evidence of two different-size male types of the same species producing different-size sperm and using different sexual positions, according to the study, published recently in the journal *BMC Evolutionary Biology.*

And there is likely an evolutionary benefit in all this, Iwata added: The sneaker sperm makes his offspring more genetically variable and thus fitter in the long run.

How Does a Sneaker Become a Sneaker?

Still a mystery, however, is how baby spear squid males become sneakers or bigger males in the first place. Scientists don't know whether sneaker status is a genetic trait passed down by parents or if environmental factors, such as water temperature or food availability, turn a baby male into a sneaker.

But research on other squid species, at least, suggests the environment may be an influence. William Gilly, a squid expert at Stanford University, said that young jumbo squid—also called Humboldt squid—completely change their life cycles during years of El Niño, a warming of tropical waters in the central and eastern Pacific Ocean.

The warmer water somehow triggers young jumbo squid to begin breeding at six months of age, when their bodies are only about 8 inches (20 centimeters) long, not including arms and tentacles. Usually the squid breed at full maturity, when their bodies are up to 3 feet (1.8 meters) long.

"The whole idea of environmental signals early in life that then cast you on some alternate trajectory," Gilly said, "may be a very common thing in squid and many animals." ∎

TRUTH:

SQUIDS SWIM BY SUCKING WATER INTO THEIR MANTLE CAVITY AND QUICKLY EXPELLING IT OUT OF A SIPHON.

Census Scopes Out
Strange New Sea Species

**Beautiful and strange:
These new sea creatures came to light
during the ten-year Census of Marine Life,
and the world is weirder for it.**

More than 6,000 new species were discovered during the Census of Marine Life, a ten-year effort to document all sea life. The project's 500-plus expeditions amassed a visual legacy as unique as the organisms uncovered—from which we selected some of the strangest.

CREATURE 1 Yeti Crab

Its fuzzy, winter-white coat might look at home in the Himalaya, but the yeti crab was discovered skittering around

hydrothermal vents about a mile and a half (2.4 kilometers) under the South Pacific off Easter Island in March 2005.

The 6-inch (15-centimeter) blind crustacean—officially *Kiwa hirsuta*—is dubbed the "yeti crab" because its long shaggy arms resemble the mythical yeti, or abominable snowman. It is such an unusual creature that a whole new family of animal had to be created to classify it.

CREATURE 2 Sea Angel

An expedition to the Arctic Ocean captured a so-called sea angel, *Clione limacina,* at about 1,148 feet (350 meters) underwater. Despite its nickname, this little angel apparently doesn't mind showing a little skin: It's actually a naked snail without a shell, scientists said in December 2009. Such marine snails—most of them the size of a lentil—are widely eaten by many species, making them the "potato chip" of the oceans, biologist Gretchen Hofmann of the University of California said in a 2008 statement.

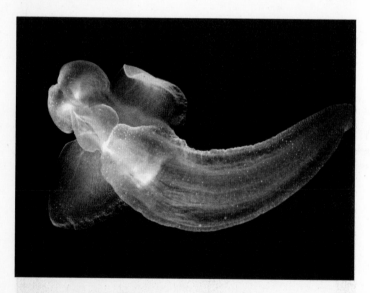

CREATURE 3 Squidworm

Squid? Worm? Initially, this new species—with bristle-based "paddles" for swimming and tentacles on its head—so perplexed Census of Marine Life researchers that they threw in the towel and simply called it squidworm.

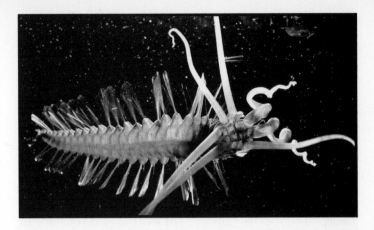

Found via a remotely operated vehicle about 1.7 miles (2.8 kilometers) under the Celebes Sea in 2007, the 4-inch-long (10-centimeter-long) creature turned out to be the first member of a new family in the Polychaeta class of segmented worms.

CREATURE 4 | Mr. Blobby

Affectionately nicknamed "Mr. Blobby," this fathead sculpin fish was discovered in 2003 in New Zealand during a

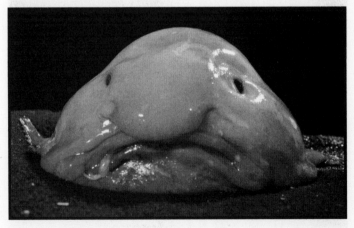

Census of Marine Life expedition, according to the Australian Museum in Sydney. Fathead sculpins—named for their large, globelike heads and floppy skin—live in the Atlantic, Indian, and Pacific Oceans at depths of between about 330 feet (100 meters) and 9,200 feet (2,800 meters). Now preserved

in 70 percent ethyl alcohol at the Australian Museum, Mr. Blobby's nose has shrunk—"and he no longer retains his 'cute' look," according to the museum's website.

| CREATURE 5 | ## Dumbo of the Deep |

Found in 2009, a deep-sea "Dumbo octopus" may look like it's all ears—but the protrusions are actually fins that help propel the animal through the darkness a mile (1.6 kilometers) under the sea. Netted during an expedition to the Mid-Atlantic Ridge, this Dumbo is among the thousands of census-documented creatures that live without

ever knowing sunlight. Reaching 6 feet (2 meters) in length and weighing 13 pounds (6 kilograms), the jumbo Dumbo species is the largest of the octopus-like creatures of the mollusk genus *Grimpoteuthis*.

CREATURE 6 | Beautiful but Deadly

"Stunningly beautiful but deadly," the Gulf of Mexico's Venus flytrap anemone acts much like its terrestrial namesake, stinging its prey with an array of tentacles, according to the U.S. Geological Survey. The species' native Gulf—along with the Mediterranean Sea, Chinese waters, the Baltic Sea, and the Caribbean Sea—are the ocean regions most under threat from human activities, according to Census of Marine Life scientists. For instance, nutrients in sewage and fertilizer washed from the land are degrading these marine habitats by creating oxygen-free "dead zones," the report says. What's more, the Gulf of Mexico oil spill may worsen these dead zones, as well as wield untold damage to the animals at the bottom of the food chain. ∎

405-Year-Old Clam

Called Longest-Lived Animal

A clam dredged from icy Arctic waters in 2007 is being hailed as the world's longest-lived animal.

Climate researchers at Bangor University in the United Kingdom recently counted 405 annual growth rings in the shells of a quahog clam. When this animal was young, Shakespeare was writing his greatest plays and the English were establishing their first settlements in the Americas.

Who Wants to Live Forever?

Here are the confirmed record holders for the world's longest lived . . .

1. **Radiated tortoise**—256 years
2. **Koi**—226 years
3. **Bowhead whale**—130 years
4. **Human being**—122 years
5. **Macaw**—106 years
6. **Elephant**—86 years
7. **Horse**—62 years
8. **Herring gull**—49 years
9. **Cat**—38 years
10. **Dog**—29 years

Counting the Rings

The team plucked the mollusk from 262-feet-deep (80-meter-deep) waters off the northern coast of Iceland. The team is studying growth lines in clam shells as part of a project to understand how the climate has changed during the past thousand years.

"On a side note, we discovered this very old clam," said Al Wanamaker, a postdoctoral researcher at the university.

Some protest the "oldest animal" designation, saying it should go to certain corals that grow together to form colonies. By this reckoning, the clam would be only the oldest non-colonial animal.

Slow Aging

Quahog clams are known for their longevity. A 220-year-old clam taken from American waters in 1982 holds the official *Guinness World Records* oldest animal title. Unofficially, the record belongs to a 374-year-old Icelandic clam housed in a German museum.

The new clam is at least 30 years older, according to the Bangor University team. The animal died when the researchers counted its rings.

"There's probably many others that are actually quite older—we just haven't found them yet," Wanamaker said. "I think in my stomach if you start getting up around 600, then maybe that would be the maximum—but that's just pure speculation," he added.

Scientists believe the secret to the clams' longevity is a slowed cell-replacement process. But why they age so slowly is unknown. "It is possible that an investigation of the tissues of these real-life Methuselahs might help us to understand the process of aging," team member Chris Richardson said in a media statement.

TRUTH: THE 405-YEAR-OLD CLAM IS NAMED MING, AFTER THE CHINESE DYNASTY.

Wanamaker added that several research teams want to study the tissue of living quahog clams to tease out the secret.

Climate Records

The researchers hope to use their shell studies to reconstruct a record of environmental changes during the past several centuries. "Just like tree rings, those growth lines vary in accordance with the environment," Wanamaker said.

Shell growth is related to water temperature, salinity, and food availability, for example. A main goal, Wanamaker said, is to determine if the climate of the last half century "is extraordinary compared to the last thousand years." ■

Photo Credits

Photograph courtesy Jerome Sueur, MNHN; 183, Courtesy Mike Picker, University of Cape Town; 184-5, John McQueen/Shutterstock; 188, kirian/Shutterstock; 190, Kees van der Krieke, Stippen.nl; 191, Mathieu B. Morin; 192 (UP), Kees van der Krieke, Stippen.nl; 192 (LO), K. van der Krieke; 194, Millard H. Sharp/Photo Researchers, Inc.; 196-7, Pitroviz/Shutterstock.

CHAPTER 5 (198-249)

198-9, Smithsonian Institution, via NASA; 200, Emory Kristof/National Geographic Stock; 202, Ozger Aybike Sarikaya/Shutterstock; 203, 6259040374/Shutterstock; 207, AP Images/Eric Gay; 209, LeCire/Wikimedia Commons; 210, Photograph from Roadrunners Internationale via Pangloss Films; 212-3, Photograph from Roadrunners Internationale via Pangloss Films; 213 (UP), Photograph from CIA via Pangloss Films; 213 (CTR), Photograph from CIA via Pangloss Films; 213 (LO), Photograph from CIA via Pangloss Films; 217, lantapix/Shutterstock; 218, Kenneth Garrett/National Geographic Stock; 223, AP Images/Matteo Borrini of Florence University, HO; 224, Steve Mann/Shutterstock; 230, James P. Blair/National Geographic Stock; 232, Oxlock/Shutterstock; 235, Oxlock/Shutterstock; 237, Adam Woolfitt/Robert Harding/Robert Harding World Imagery/Getty Images; 238-9, John Young/Shutterstock; 240, NASA; 241, NASA; 242 (UP), NASA; 242 (LO), NASA; 243, NASA; 244, NASA; 245, NASA; 246, Reuters/Corbis; 248-9, R. Formidable/Shutterstock.

CHAPTER 6 (250-285)

250-1, David Trood/The Image Bank/Getty Images; 253, John Downer/Oxford Scientific/Getty Images; 256, Frans Lanting/National Geographic Stock; 258, Albo003/Shutterstock; 259, The Roslin Institute at the University of Edinburgh; 261, FLPA/Alamy; 264-5, Potapov Alexander/Shutterstock; 267, Silvia Reiche/FN/Minden Pictures/National Geographic Stock; 270-1, Oleg Iatsun/Shutterstock; 272, Rolf Nussbaumer Photography/Alamy; 275, Koshevnyk/Shutterstock; 276, Courtesy Fundación ProAves; 281, Jon Reed/National Geographic My Shot; 282, silver tiger/Shutterstock; 285, Supplied by WENN.com, via Newscom.

CHAPTER 7 (286-333)

286-7, NASA/JPL-Caltech; 288, NASA; 290, Michael Monahan/Shutterstock; 291, ESO/M. Kornmesser; 293, John T Takai/Shutterstock; 294, STEREO/NASA; 299, ESA/AOES Medialab; 301, web-artist/Shutterstock; 302-3, Paul Schenk, LPI; 303 (UP), SSI/NASA; 303 (CTR UP), Paul Schenk, LPI; 303 (CTR LO), SSI/NASA; 303 (LO), SSI/NASA; 306, NASA; 309, NASA, ESA, and M. Showalter (SETI Institute); 313, FreeSoulProduction/Shutterstock; 317, Illustration courtesy Swinburne Astronomy Productions; 319, NASA, ESA, the Hubble Heritage (STScI/AURA)-ESA/Hubble Collaboration, and A. Evans (University of Virginia, Charlottesville/NRAO/Stony Brook University); 320-1, nex999/Shutterstock; 325, David A. Aguilar (CfA), TrES, Kepler, NASA; 329, NASA.

CHAPTER 8 (334-393)

334-5, mareandmare/Shutterstock; 337, Morphart Creations Inc./Shutterstock; 338, Habakkuk Commentary, columns 5-8, Qumran Cave 1, 1st century BC (parchment)/The Israel Museum, Jerusalem, Israel/The Bridgeman Art Library; 340, Levent AVCI/Shutterstock; 343, Bojanovic/Shutterstock; 345, Dimedrol68/Shutterstock; 346-7, Binkski/Shutterstock; 348, Morphart Creations Inc./Shutterstock; 351, joSon/Stone/Getty Images; 352, Raymond Gehman/National Geographic Stock; 353, Travelpix Ltd/Photographer's Choice/Getty Images; 354, Ralf Siemieniec/Shutterstock; 355, agap/Shutterstock; 356, Curtis Kautzer/Shutterstock; 357, Pius Lee/Shutterstock; 360, pio3/Shutterstock; 364, Courtesy José C. Jiménez López; 368, Michael Lavin, courtesy Preservation Virginia; 370, Roman Sotola/Shutterstock; 372, Hanna J/Shutterstock; 375, The Black Death, 1348 (engraving) (b&w photo), English School, (14th century)/Private Collection/The Bridgeman Art Library; 378, Christian Mundigler/AFP/Getty Images/Newscom; 381, Don Maitz; 382 (LE), Courtesy Wendy M. Welsh, North Carolina Department of Cultural Resources; 382 (RT), Photograph courtesy Wendy M. Welsh, North Carolina Department of Cultural Resources; 382-3, Courtney Platt/National Geographic; 383, Photograph by Shanna Daniel, North Carolina Department of Cultural Resources; 386, Menahem Kahana/AFP/Getty Images/Newscom; 388-9, Algol/Shutterstock; 391,

Fernbank Museum of Natural History; 393, Fernbank Museum of Natural History.

CHAPTER 9 (394-435)

394-5, Carsten Peter/Speleoresearch & Films/National Geographic Stock; 396, Photograph courtesy of Leif Kullman at Umeå University's Department of Ecology and Environmental Science; 398, Emir Simsek/Shutterstock; 400, Everett Collection/Shutterstock; 402, Vesa Andrei Bogdan/Shutterstock; 405, Weatherthings.com; 406, CataVic/Shutterstock; 408, Courtesy Paulo Raquec; 411, Mark Thiessen, NGP; 412-3, iraladybird/Shutterstock; 417, Carsten Peter/Speleoresearch & Films/National Geographic Stock; 422, SimonasP/Shutterstock; 424, Jose Dominguez/AFP/Getty Images; 424-5, Photograph courtesy Jason deCaires Taylor; 425 (UP), Photograph courtesy Jason deCaires Taylor; 425 (CTR), AP Photo/Miguel Tovar; 425 (LO), Photograph courtesy Jason deCaires Taylor; 429, Morphart Creations Inc./Shutterstock; 431, Carsten Peter/National Geographic Stock; 433, Emir Simsek/Shutterstock; 434, Cameron McIntire.

CHAPTER 10 (436-479)

436-7, Damnfx/National Geographic Stock; 439, Tairy Greene/Shutterstock; 442, Chris Ison/PA Wire URN:7967528 (Press Association via AP Images); 442-3, Courtesy Jurassic Coast Team, Dorset County Council; 443 (UP), Courtesy Jurassic Coast Team, Dorset County Council; 443 (LO), Raul Martin/National Geographic Stock; 444, Mackey Creations/Shutterstock; 446, Jason Bourque; 446-7, Marie Appert/Shutterstock; 449, Stubblefield Photography/Shutterstock; 451, Guillermo W. Rougier; 453, Leremy/Dreamstime.com; 455, Illustration courtesy Francisco Gascó, Mike Taylor, and Matt Wedel; 457, Antonuk/Shutterstock; 459, Morphart Creations Inc./Shutterstock; 461, Henry P. Tsai, University of Missouri; 462, pio3/Shutterstock; 464, Graphic by Katie Parker, National Geographic Digital Media; 465, Koshevnyk/Shutterstock; 467, Katie Parker, National Geographic Digital Media; 468, Joe McNally/National Geographic Stock; 472, Courtesy Mike Hettwer; 472-3, Courtesy Ricardo Martinez; 473 (UP), Mike Hettwer/Eureka/Handout/picture alliance/dpa/Newscom; 473 (LO), Courtesy Mike Hettwer; 477, O. Louis Mazzatenta/National Geographic Stock; 478-9, Vule/Shutterstock.

CHAPTER 11 (480-531)

480-1, Colin Parker/National Geographic My Shot; 483, Pisces Sportfishing Fleet/Rex Features, For more information visit http://www.rexfeatures.com/stacklink/VRUEOJKZM (Rex Features via AP Images); 486, Karen Gowlett-Holmes/Supplied by WENN.com, via Newscom; 487 (UP), CSIRO/Supplied by WENN.com, via Newscom; 487 (LO), Courtesy CSIRO; 488, Courtesy CSIRO; 489, Photograph courtesy Scott Gardner; 490, abrakadabra/Shutterstock; 493, Denis Barbulat/Shutterstock; 495, beta757/Shutterstock; 497, C & M Fallows/SeaPics.com; 498, Albo003/Shutterstock; 500, SFerdon/Shutterstock; 503, David Wachenfeld/Triggerfish Images; 505, Roger Steene/Supplied by WENN.com, via Newscom; 506, Thirteen-Fifty/Shutterstock; 509, Courtesy Bangor Daily News/Abigail Curtis; 510, Monterey Bay Aquarium Research Institute; 512, Greenland Institute of Natural Resources; 513, Greenland Institute of Natural Resources; 514 (UP), Henrik Carl, Natural History Museum of Denmark; 514 (LO), Greenland Institute of Natural Resources; 515, Henrik Carl, Natural History Museum of Denmark; 516, Suthep Kritsanavarin; 519, Photograph courtesy Don Demaria; 521, Anton Novik/Shutterstock; 525, Courtesy A. Fifis, IFREMER; 526, Courtesy Kevin Raskoff, Monterey Peninsula College; 527 (UP), Courtesy Laurence Madin, Woods Hole Oceanographic Institution; 527 (LO), Courtesy Kerryn Parkingson, NORFANZ; 528, Courtesy David Shale; 529, Courtesy Ian MacDonald, Florida State University; 530, Petrafler/Shutterstock.

Index

Boldface indicates illustrations.